D0587368

Student Support Materials for
Edexcel A Level Maths
Core 3

and Sue Langham

9112000270867

William Collins' dream of knowledge for all began with the publication of his first book in 1819. A self-educated mill worker, he not only enriched millions of lives, but also founded a flourishing publishing house. Today, staying true to this spirit, Collins books are packed with inspiration, innovation and practical expertise. They place you at the centre of a world of possibility and give you exactly what you need to explore it.

Collins. Freedom to teach.

Published by Collins
An imprint of HarperCollins *Publishers*
77–85 Fulham Palace Road
Hammersmith
London
W6 8JB

© HarperCollins*Publishers* Limited 2012

10 9 8 7 6 5 4 3 2 1

ISBN-13: 978-0-00-747603-9

British Library Cataloguing in Publication Data. A Catalogue record for this publication is available from the British Library.

Commissioned by Lindsey Charles and Emma Braithwaite
Project managed by Lindsey Charles
Edited and proofread by Susan Gardner
Reviewed by Stewart Townend
Design and typesetting by Jouve India Private Limited
Illustrations by Ann Paganuzzi
Index compiled by Michael Forder
Cover design by Angela English
Production by Simon Moore

Printed and bound in Spain by Graficas Estella

This material has been endorsed by Edexcel and offers high quality support for the delivery of Edexcel qualifications.

Edexcel endorsement does not mean that this material is essential to achieve any Edexcel qualification, nor does it mean that this is the only suitable material available to support any Edexcel qualification. No endorsed material will be used verbatim in setting any Edexcel examination and any resource lists produced by Edexcel shall include this and other appropriate texts. While this material has been through an Edexcel quality assurance process, all responsibility for the content remains with the publisher.

Copies of official specifications for all Edexcel qualifications may be found on the Edexcel website - www.edexcel.com

Browse the complete Collins catalogue at: www.collinseducation.com

Acknowledgements
The publishers wish to thank the following for permission to reproduce photographs. Every effort has been made to trace copyright holders and to obtain their permission for the use of copyright material. The publishers will gladly receive any information enabling them to rectify any error or omission at the first opportunity.

Cover image: Abstract building reflection © Jahina001 | Dreamstime.com

MIX
Paper from
responsible sources

FSC
www.fsc.org **FSC™ C007454**

FSC™ is a non-profit international organisation established to promote the responsible management of the world's forests. Products carrying the FSC label are independently certified to assure consumers that they come from forests that are managed to meet the social, economic and ecological needs of present and future generations, and other controlled sources.

Find out more about HarperCollins and the environment at
www.harpercollins.co.uk/green

Welcome to Collins Student Support Materials for Edexcel A level Mathematics. This page introduces you to the key features of the book which will help you to succeed in your examinations and to enjoy your maths course.

The chapters are organised by the main sections within the specification for easy reference. Each one gives a succinct explanation of the key ideas you need to know.

Examples and answers

After ideas have been explained the worked examples in the green boxes demonstrate how to use them to solve mathematical problems.

Method notes

These appear alongside some of the examples to give more detailed help and advice about working out the answers.

Essential ideas

These are other ideas which you will find useful or need to recall from previous study.

Exam tips

These tell you what you will be expected to do, or not to do, in the examination.

Stop and think

The stop and think sections present problems and questions to help you reflect on what you have just been reading. They are not straightforward practice questions - you have to think carefully to answer them!

Practice examination section

At the end of the book you will find a section of practice examination questions which help you prepare for the ones in the examination itself. Answers with full workings out are provided to all the examination questions so that you can see exactly where you are getting things wrong or right!

Notation and formulae

The notation and formulae used in this examination module are listed at the end of the book just before the index for easy reference. The formulae list shows both what you need to know and what you will be given in the exam.

Contents

Contents

In Core 2 you studied the properties of polynomials. In particular you were introduced to the factor and remainder theorems and learnt how to divide polynomials. We begin Core 3 with a review of the techniques for simplifying rational polynomial expressions.

Simplifying algebraic fractions

Algebraic division

Essential notes

A rational polynomial expression is one in which one polynomial is divided by another polynomial. It is also called an algebraic fraction.

Example

Simplify the following algebraic fractions:

a) $\dfrac{5x + 5}{2x + 2}$ b) $\dfrac{a^3}{a^2 + 2a}$

c) $\dfrac{x^2 - 3x}{x^2 - 9}$ d) $\dfrac{x^2 + 6x + 5}{x^2 - x - 2}$

Method notes

Start by factorising the numerator and denominator of the fraction. Then divide through by any factors which are common to the numerator and the denominator.

In (a) $(x + 1)$ is a common factor.

In (b), you can cancel a as it is a common factor to give $\dfrac{a^2}{a + 2}$ then you cannot cancel a further as it is no longer a common factor.

In (c) and (d) similarly, you cannot cancel out x further in $\dfrac{x}{x + 3}$ and $\dfrac{x + 5}{x - 2}$.

Answer

a) $\dfrac{5x + 5}{2x + 2} = \dfrac{5(x + 1)}{2(x + 1)} = \dfrac{5}{2}$

b) $\dfrac{a^3}{a^2 + 2a} = \dfrac{a^3}{a(a + 2)} = \dfrac{a^2}{(a + 2)}$

c) $\dfrac{x^2 - 3x}{x^2 - 9} = \dfrac{x(x - 3)}{(x + 3)(x - 3)} = \dfrac{x}{(x + 3)}$

d) $\dfrac{x^2 + 6x + 5}{x^2 - x - 2} = \dfrac{(x + 1)(x + 5)}{(x + 1)(x - 2)} = \dfrac{(x + 5)}{(x - 2)}$

Rational functions with linear denominators

We can simplify such algebraic fractions by using the **remainder theorem**.

The remainder theorem states that:

If a polynomial $p(x)$ is divided by $(ax - b)$ then the remainder is $p\left(\dfrac{b}{a}\right)$

where a and b are constants.

Essential notes

The remainder theorem and factor theorem for polynomials were introduced in Core 2.

$(ax - b)$ is a linear function.

Example

When $f(x) = x^3 + kx^2 - 7x + 3$ is divided by $(x - 3)$ the remainder is 18. By applying the remainder theorem,

a) find the value of k

b) write $f(x)$ in terms of a quadratic polynomial.

Answer

a) **Step 1**: Given $f(x) = x^3 + kx^2 - 7x + 3$ the remainder when $f(x)$ is divided by $(x - 3)$ is:

$f(3) = 27 + 9k - 21 + 3 = 9 + 9k$

b) **Step 2**: Given that the remainder is 18, $9 + 9k = 18$ therefore $k = 1$

Substituting $k = 1$ from step 2:

$f(x) = x^3 + x^2 - 7x + 3$

So we can write $x^3 + x^2 - 7x + 3 = g(x)(x - 3) + 18$ where $g(x)$ is a quadratic polynomial.

Method notes

b) In an arithmetic example, when 17 is divided by 2 the remainder is 1. This means that we can write $17 = 8 \times (2) + 1$

In the next example we see how to find the form of the quadratic $g(x)$.

Example

a) Use the remainder theorem find the remainder when $x^3 + 2x^2 + 5x - 2$ is divided by $(x - 1)$.

b) Hence write $x^3 + 2x^2 + 5x - 2$ in terms of a quadratic polynomial $(ax^2 + bx + c)$ and find the values of a, b and c.

Answer

a) The remainder when $f(x) = x^3 + 2x^2 + 5x - 2$ is divided by

$(x - 1)$ is $f(1) = 1 + 2 + 5 - 2 = 6$

b) We can write $f(x)$ as

$x^3 + 2x^2 + 5x - 2 \equiv (\text{quadratic}) \times (x - 1) + 6$

$x^3 + 2x^2 + 5x - 2 \equiv (ax^2 + bx + c)(x - 1) + 6$ \hfill (1)

and this is true for all values of x.

Step 1: Let $x = 0$ in (1) $\Rightarrow -2 = c \times (-1) + 6 \Rightarrow c = 8$

Step 2: Comparing the coefficients of x^3 in (1) gives $1 = a$

Step 3: Comparing the coefficients of x^2 in (1) gives
$2 = -a + b$

Step 4: Substituting $a = 1$ from step 2 in the result from step 3 gives $b = 3$

This means that $x^3 + 2x^2 + 5x - 2 \equiv (1x^2 + 3x + 8)(x - 1) + 6$

Method notes

$ax^2 + bx + c$ is the general form of a quadratic.

The last two examples show that we can rewrite a polynomial $f(x)$ as:

$$f(x) \equiv g(x)\,p(x) + D$$

where $p(x)$ is a linear expression, $g(x)$ is a polynomial of degree one less than the given function $f(x)$ and D is the remainder.

If we now divide every term of

$x^3 + 2x^2 + 5x - 2 \equiv (1x^2 + 3x + 8)(x - 1) + 6$ by $(x - 1)$ this gives

$$\frac{x^3 + 2x^2 + 5x - 2}{x - 1} = x^2 + 3x + 8 + \frac{6}{x - 1}$$

In this fraction the degree of the numerator is higher than that of the denominator and this is called an **improper fraction of polynomials**.

Essential notes

The 'degree' is the highest power of the polynomial. A cubic polynomial has degree 3 and a quadratic polynomial has degree 2.

We can use the remainder theorem to change it into a quadratic polynomial and an algebraic **proper fraction**:

$$x^2 + 3x + 8 + \frac{6}{(x-1)}$$

In a proper fraction the degree of the numerator is less than the degree of the denominator.

An alternative approach for simplifying improper algebraic fractions uses the method of long division of polynomials.

Essential notes

Long division of polynomials was covered in Core 2.

Method notes

In long division examples you stop dividing when the remainder (here it is 6 and is a constant) is of lower degree (or power) than the divisor. In this case the divisor which is $(x + 1)$ is a linear function of degree 1.

Example

Use long division to divide $x^3 + 2x^2 + 5x - 2$ by $(x - 1)$.

Answer

$$
\begin{array}{r}
x^2 + 3x + 8 \\
(x-1)\overline{)x^3 + 2x^2 + 5x - 2} \\
\underline{x^3 - x^2} \\
3x^2 + 5x - 2 \\
\underline{3x^2 - 3x} \\
8x - 2 \\
\underline{8x - 8} \\
6
\end{array}
$$

$$\frac{x^3 + 2x^2 + 5x - 2}{(x-1)} = x^2 + 3x + 8 + \frac{6}{(x-1)}$$

Rational functions with quadratic denominators

Example

Use long division to write the fraction

$$\frac{x^4 + x^3 + 2x^2 - 3x + 4}{x^2 + x + 1}$$

as a quadratic function plus a proper algebraic fraction.

Answer

$$
\begin{array}{r}
x^2 + 1 \\
(x^2 + x + 1)\overline{)x^4 + x^3 + 2x^2 - 3x + 4} \\
\underline{x^4 + x^3 + x^2} \\
x^2 - 3x + 4 \\
\underline{x^2 + x + 1} \\
- 4x + 3
\end{array}
$$

← degree of remainder < degree of divisor

Method notes

In this example you stop dividing out because the degree of the remainder $(-4x + 3)$ is less than the degree of the divisor $(x^2 + x + 1)$.

Therefore $\dfrac{x^4 + x^3 + 2x^2 - 3x + 4}{(x^2 + x + 1)} = x^2 + 1 + \dfrac{(-4x + 3)}{(x^2 + x + 1)}$

$x^2 + 1$ is the quadratic function and $\dfrac{(-4x + 3)}{(x^2 + x + 1)}$ is the proper algebraic fraction.

Adding algebraic fractions

When adding numerical fractions such as $\frac{2}{3} + \frac{1}{4}$ the first step is to write each fraction in terms of a common denominator which is the lowest common multiple (LCM) of the denominators in the given fractions.

Here the lowest common multiple of 3 and 4 is 12. So

$$\frac{2}{3} + \frac{1}{4} = \frac{8}{12} + \frac{3}{12} = \frac{11}{12}$$

We apply the same technique when adding algebraic fractions.

Example

Simplify the following fractions:

a) $\frac{3}{x} + \frac{4}{x^2}$

b) $\frac{3}{(x-1)} + \frac{2}{(x+1)}$

c) $\frac{4}{(x^2+3x+2)} - \frac{3}{(x^2+4x+4)}$

Answer

a) $\frac{3}{x} + \frac{4}{x^2} \equiv \frac{3x}{x^2} + \frac{4}{x^2} \equiv \frac{3x+4}{x^2}$

b) $\frac{3}{(x-1)} + \frac{2}{(x+1)} \equiv \frac{3(x+1)}{(x-1)(x+1)} + \frac{2(x-1)}{(x-1)(x+1)}$

$$\equiv \frac{5x+1}{(x-1)(x+1)}$$

c) $\frac{4}{(x^2+3x+2)} - \frac{3}{(x^2+4x+4)} \equiv \frac{4}{(x+2)(x+1)} - \frac{3}{(x+2)^2}$

$$\equiv \frac{4(x+2)}{(x+1)(x+2)^2} - \frac{3(x+1)}{(x+1)(x+2)^2}$$

$$\equiv \frac{x+5}{(x+1)(x+2)^2}$$

Functions

You have met five key functions in Core 1 and Core 2:

- a **linear function** such as $f(x) = 5x - 7$
- a **quadratic function** such as $f(x) = x^2 - 4x - 3$
- a **cubic function** such as $f(x) = x^3 + 2x + 4$
- a **trigonometric function** such as $f(x) = \sin x$
- a **logarithmic function** such as $f(x) = \log_{10} x$

The first three are special cases of **polynomial functions**.

Each of the functions given above, except the logarithmic one, is defined for any real value of x. There is only **one** y-value for each x-value. This can be shown clearly by drawing a graph of $y = f(x)$.

Essential notes

A common denominator means that each fraction is written with the same denominator.

The lowest common multiple (LCM) of two numbers is the smallest number which both numbers will divide into to give a whole number answer.

Method notes

a) The LCM of x and x^2 is x^2 because $\frac{x^2}{x} = x$ and $\frac{x^2}{x^2} = 1$

b) The LCM of $(x-1)$ and $(x+1)$ is $(x-1) \times (x+1)$ as it can be divided by $(x-1)$ and $(x+1)$ to give a non-fractional answer.

c) The first step is to factorise the denominators:
$x^2 + 3x + 2 = (x+2)(x+1)$
$x^2 + 4x + 4 = (x+2)^2$

The LCM of $(x+2)(x+1)$ and $(x+2)^2$ is $(x+2)^2(x+1)$ because this can be divided by both denominators to give a non-fractional answer.

Essential notes

A real number is any rational number (which can be expressed as a ratio of any two integers) or an irrational number (which cannot be expressed as a ratio of any two integers). An integer is a whole number.

However, the reverse is not always true and there is sometimes more than one x-value for a y-value. For example if:

$$f(x) = x^2 - 4x - 3 = 2$$
$$y = x^2 - 4x - 3 = 2$$

rewriting gives $f(x) = x^2 - 4x - 5 = 0$

factorising gives $f(x) = (x + 1)(x - 5) = 0$

So there are two solutions for x: $x = -1$ and $x = 5$

Logarithmic functions are not defined for all real values of x. The example of a logarithmic function such as $f(x) = \log_{10}x$ is only defined for $x > 0$

The graphs of the five functions (figures 1.1 to 1.5) illustrate some of their important properties.

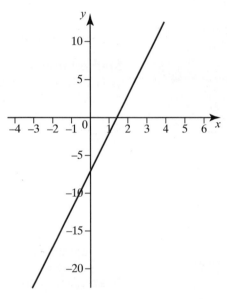

Fig.1.1
Graph of the linear function $y = 5x - 7$

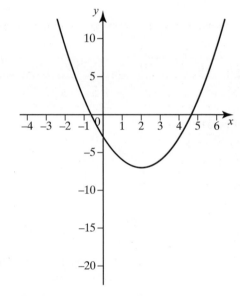

Fig.1.2
Graph of the quadratic function $y = x^2 - 4x - 3$

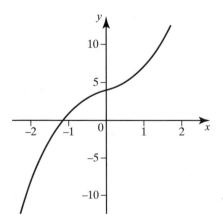

Fig.1.3
Graph of the cubic function $y = x^3 + 2x + 4$

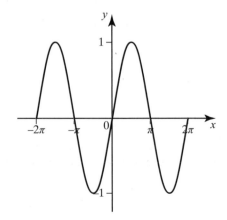

Fig.1.4
Graph of the trigonometric function $y = \sin x$.

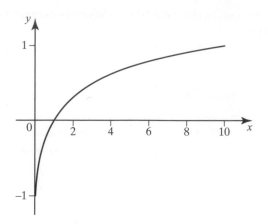

Fig 1.5
Graph of logarithmic function
$y = \log_{10} x$.

By looking at these graphs you can deduce that there may be some restrictions on the x- and y- values as can be seen in the table below:

function	x-values	y-values
$f(x) = 5x - 7$	$\{x: -\infty < x < \infty\}$	$\{y: -\infty < y < \infty\}$
$f(x) = x^2 - 4x - 3$	$\{x: -\infty < x < \infty\}$	$\{y: -7 \le y < \infty\}$
$f(x) = x^3 + 2x + 4$	$\{x: -\infty < x < \infty\}$	$\{y: -\infty < y < \infty\}$
$f(x) = \sin x$	$\{x: -\infty < x < \infty\}$	$\{y: -1 \le y \le 1\}$
$f(x) = \log_{10} x$	$\{x: x > 0\}$	$\{y: -\infty < y < \infty\}$

Mapping

A **mapping** is a relationship or rule which operates from one set of elements to another set of elements.

A **function** is a mapping of elements from a set A, called the **domain,** to elements in a set B, called the **range**.

In functions the domain is the set of x-values and the range is the set of y-values to which they are mapped. For each element in the domain under this mapping, there is a unique element in the range.

Each element in the domain is an **object** which has a corresponding element in the range called an **image**.

The functions:

$$f(x) = 5x - 7 \text{ (linear)}$$

$$g(x) = x^3 \text{ (cubic)}$$

and $h(x) = \log_{10} x$ (logarithmic)

are examples of **one-one functions** for which each element value in the range has a unique element value in the domain.

Essential notes

Set notation:

$\{x : 1, 2, 3, 4\}$ means the **set** of numbers 1, 2, 3, 4 and each member of the set is called an **element**.

Essential notes

Unique means only one.

We abbreviate the domain set to 'domain' and the range set to 'range'.

Essential notes

You can use any letter for a function e.g. $f(x)$, $g(x)$, $h(x)$ all represent **different** functions of x.

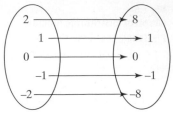

Domain (x-values)   Range (y-values)

Fig. 1.6
The one-one function $y = f(x) = x^3$.

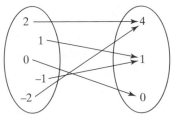

Domain (x-values)   Range (y-values)

Fig. 1.7
$y = f(x) = x^2$ many-one function.

Essential notes

Functions must be either one-one or many-one mappings.

Essential notes

$y = f(x)$ means we can state the range as:

$\{y: -1, 1, 3, 5, 7\}$ or

$\{f(x): -1, 1, 3, 5, 7\}$.

The mapping diagram in Fig. 1.6 illustrates this idea for the one-one function $y = f(x) = x^3$.

The functions:

$f(x) = x^2$ (quadratic)

and $f(x) = \sin x$ (trigonometric)

are examples of **many-one functions** for which element values in the range may have more than one element value in the domain, for example:

if $f(x) = x^2$ then $f(2) = 4$ and $f(-2) = 4$

The mapping diagram in Figure 1.7 illustrate this idea for the many-one function $y = f(x) = x^2$.

Example

For each of the following functions, with the given domain,

   i) write down the range,

   ii) state whether the function is one-one or many-one.

a) $f(x) = x + 3$ $\{x: -4, -2, 0, 2, 4\}$

b) $g(x) = \dfrac{2}{x^2}$ $\{x: -2, -1, 1, 2, 3\}$

Answer

a) i) $f(x) = x + 3$ has a domain $\{x: -4, -2, 0, 2, 4\}$

Work out the function values for each element of the domain and these will be the elements of the range.

Step 1: $x = -4 \Rightarrow f(-4) = -4 + 3 = -1$

Step 2: $x = -2 \Rightarrow f(-2) = -2 + 3 = 1$

Step 3: $x = 0 \Rightarrow f(0) = 0 + 3 = 3$

Step 4: $x = 2 \Rightarrow f(2) = 2 + 3 = 5$

Step 5: $x = 4 \Rightarrow f(4) = 4 + 3 = 7$

Therefore the range is $\{y: -1, 1, 3, 5, 7\}$.

Each of the x-values has given only one $f(x)$ or y-value which means that the function is one-one.

b) $g(x) = \dfrac{2}{x^2}$ has a domain $\{x: -2, -1, 1, 2, 3\}$.

i) The method is the same as in (a) and the results show that:

$g(-2) = g(2) = \dfrac{1}{2}: f(-1) = f(1) = 2: f(3) = \dfrac{2}{9}$

We can now state that the range is $\left\{y: \dfrac{1}{2}, 2, \dfrac{2}{9}\right\}$

ii) Two x-values have given the same $f(x)$ or y-value so the function is many-one.

Function notations

There are two different function notations. So far in this chapter we have written the function as $f(x) = y$ for which x is an element (or object) in the domain of the function and y is an element (or image) in the range of the function.

An alternative notation is $f: x \mapsto y$.

For example $f(x) = x + 3$ with domain $\{x: -4, -2, 0, 2, 4\}$ can also be written as $f: x \mapsto x + 3$ with domain $\{x: -4, -2, 0, 2, 4\}$.

The following examples illustrate the use of this notation.

Example

For each of the following functions, with the given domain, where x is a real number

   i)   sketch a graph of the function,

   ii) write down the range,

   iii) state whether the function is one-one or many-one.

a) $f: x \mapsto x^2 - 5x + 6$ $\{x: x \geq 0\}$

b) $f: x \mapsto x^3 - 12x$ $\{x: -3 \leq x \leq 3\}$

c) $f: x \mapsto \dfrac{1}{x+1}$ $\{x: -2 \leq x \leq 2, x \neq -1\}$

Answer

a) i) Given $f: x \mapsto x^2 - 5x + 6$ $\{x: x \geq 0\}$ means that

   $y = x^2 - 5x + 6$ with domain the set of all real numbers ≥ 0

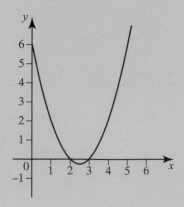

  ii) Minimum value of y is $-\dfrac{1}{4}$ so the range is $\left\{y: y \geq -\dfrac{1}{4}\right\}$.

  iii) Points $(2, 0)$ and $(3, 0)$ show that two x-values give the same y-value so the function is many-one.

b) i) Given $f: x \mapsto x^3 - 12x$ $\{x: -3 \leq x \leq 3\}$ means that $y = x^3 - 12x$ with domain the set of all real numbers from -3 to $+3$

Method notes

i) To sketch the graph find where the graph crosses the x-axis i.e. where $y = 0$

  Since $y = x^2 - 5x + 6$ the curve crosses the x axis where $x^2 - 5x + 6 = 0$

  Factorising gives

  $(x - 3)(x - 2) = 0$ so $x = 2$ or $x = 3$

  Therefore the curve crosses the x-axis at the points $(2, 0)$ and $(3, 0)$.

ii) By symmetry the minimum turning point occurs midway between these two points where $x = 2.5$ and so $y = -\dfrac{1}{4}$

Fig. 1.8
Graph of $y = x^2 - 5x + 6$ $\{x: x \geq 0\}$

Fig. 1.9
Graph of $y = x^3 - 12x$ {$x: -3 \leq x \leq 3$}.

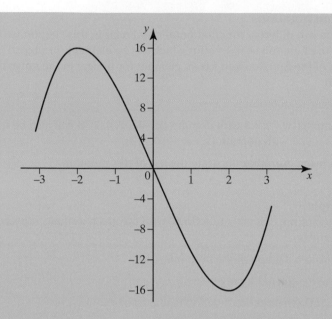

Method notes

i) To sketch the graph find where it crosses the x-axis i.e. where

$y = 0: y = x^3 - 12x = 0$

Factorising gives $x(x^2 - 12) = 0$ so $x = 0$ or $x = \pm\sqrt{12}$

But only $x = 0$ is in the given domain {$x: -3 \leq x \leq 3$} as $\pm\sqrt{12} = \pm 3.464$

ii) The turning points are where

$\dfrac{dy}{dx} = 0$

ii) Maximum and minimum values of y are16 and -16 so the range is {$y: -16 \leq y \leq 16$}.

iii) The graph shows that by taking points either side of the turning points, two different x-values will give the same y-value therefore the function is many-one.

If $y = x^3 - 12x$

$\dfrac{dy}{dx} = 3x^2 - 12$

so $0 = 3x^2 - 12 = 3(x^2 - 4)$

giving $0 = 3(x + 2)(x - 2)$ so $x = -2$ or $x = +2$

If $x = -2$: $y = (-2)^3 - 12(-2) = -16$

If $x = 2$: $y = (2)^3 - 12(2) = 16$

Hence the turning points are $(-2, 16)$ and $(2, -16)$. Given the shape of cubic curves (Core 2) we know that $(-2, 16)$ is a maximum turning point and $(2, -16)$ is a minimum turning point.

c) i) $x \mapsto \dfrac{1}{x + 1}$ {$x: -2 \leq x \leq 2, x \neq -1$}

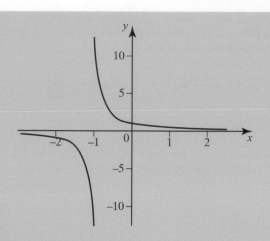

Fig. 1.10
Graph of $y = \dfrac{1}{x + 1}$

$\{x: -2 \le x \le 2, x \ne -1\}$

ii) The domain is given as $\{x: -2 \le x \le 2, x \ne -1\}$

If $x = 2$ then $y = \dfrac{1}{x + 1} = \dfrac{1}{2 + 1} = \dfrac{1}{3}$ which means that above the

x-axis the smallest value of $y = \dfrac{1}{3}$

If $x = -2$ then $y = \dfrac{1}{x + 1} = \dfrac{1}{-2 + 1} = -1$ which means that below

the x-axis the greatest value of y for the given domain $= -1$

therefore the range is $\{y: -\infty \le y \le -1, \dfrac{1}{3} \le y \le \infty\}$.

iii) From the graph the function is one-one for the domain specified, the function being undefined for $x = 1$

Method notes

i) To sketch the graph of $y = \dfrac{1}{x + 1}$ it is helpful to know the shape of the basic function $y = \dfrac{1}{x}$.

We can then see from the work covered in Core 1 that $y = \dfrac{1}{x + 1}$ is a translation of the graph $y = \dfrac{1}{x}$, 1 unit to the left, parallel to the x-axis.

The set of real numbers has the special notation $\Re$. $x \in \Re$ means x is any member of the set of real numbers.

$\Re^+$ means the set of positive real numbers and $\Re^-$ means the set of negative real numbers.

Composite functions

Consider the two functions $f(x) = 3x + 1$ and $g(x) = x^2$ for which each function has domain $\Re$. We now consider different ways of combining these two functions of x.

Suppose that we apply the function $f(x)$ to the object $x = 2$ from the domain then the image of $x = 2$ is $f(2) = 3 \times 2 + 1 = 7$

If we now apply the function $g(x)$ to the image of $f(x)$ i.e. 7 then the image is $g(7) = 7^2 = 49$

This process is illustrated in the diagram on the right.

Repeating this process for any value of x in the domain of $f(x)$ is shown in Figure 1.12.

This process is an example of a **composite function** written as $gf(x)$.

As is illustrated above for $x = 2$, $gf(2) = 49$ and for any $x \in \Re$, $gf(x) = (3x + 1)^2$ for this composite function.

Exam tips

You must know what the $\Re$ notation means as it may be used in the examination.

Fig.1.11
The composite function $gf(2)$.

$x \xrightarrow{\;f(x)\;} \boxed{3x + 1} \xrightarrow{\;g(x)\;} \boxed{(3x + 1)^2}$

Fig. 1.12
The composite function $gf(x)$.

Essential notes

For composite functions deal with the function written next to the 'x' first so for $gf(x)$ apply the function rule 'f' first then apply the function rule 'g'.

For $fg(x)$ apply the function rule 'g' first then apply the function rule 'f'.

Essential notes

In general $fg(x) \neq gf(x)$.

Essential notes

$ff(x)$ is often written as $f^2(x)$. Similarly $gg(x) \equiv g^2(x)$.

Method notes

It may help to introduce a new variable, say u, into your working.

Think of $g(x) = u = 2x^2 + x$

Then $f(g(x)) = f(u) = u^2 + 1$

Now substitute $2x^2 + x$ for u to give $f(g(x)) = (2x^2 + x)^2 + 1$

The order in which the functions are applied is important. If you carry out the function g first and then carry out the function f, you get the composite function $fg(x)$ illustrated in Figure 1.13.

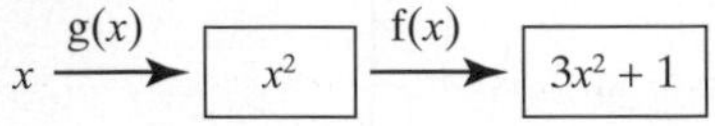

$$x \xrightarrow{\ g(x)\ } \boxed{x^2} \xrightarrow{\ f(x)\ } \boxed{3x^2 + 1}$$

Fig. 1.13
The composite function $fg(x)$.

For any $x \in \Re$, $fg(x) = 3x^2 + 1$ for these two functions showing that in this example $fg(x)$ and $gf(x)$ are not equal.

Stop and think 1

Is it ever true that for the functions $f(x)$ and $g(x)$ that $fg(x) = gf(x)$ where $x \in \Re$?

Give reasons for your answer.

Example

Functions f and g are defined by: $f(x) = x^2 + 1$ and $g(x) = 2x^2 + x$. Find expressions for

a) $fg(x)$ b) $gf(x)$

c) $ff(x)$ d) $gg(x)$

Answer

a) Given $f(x) = x^2 + 1$ and $g(x) = 2x^2 + x$, to work out $fg(x)$ we apply the 'g' function rule first then the 'f' function rule.

Step 1: State the 'g' function rule $\Rightarrow g(x) = (2x^2 + x)$

Step 2: Apply the 'f' function rule to the result $(2x^2 + x)$ in step 1 to give $f((g(x))$.

The 'f' function states that you square whatever is the object (now $(2x^2 + x)$) and add on 1 which gives $f(g(x)) = (2x^2 + x)^2 + 1$

Step 3: Simplify the equation:

$$f(g(x)) = 4x^4 + 4x^3 + x^2 + 1$$

b) To work out $gf(x)$ we apply the 'f' rule first then the 'g' rule.

Step 1: State the 'f' rule $\Rightarrow f(x) = x^2 + 1$

Step 2: Apply the 'g' rule to the result $(x^2 + 1)$ in step 1 to give $g(f(x))$.

The 'g' rule states that you square whatever is the object, (now $(x^2 + 1)$) multiply by 2 then add on the object $\Rightarrow g((f(x)) = g(x^2 + 1)$
$= 2(x^2 + 1)^2 + (x^2 + 1)$

Step 3: Simplify the equation:

$$g(f(x)) = 2(x^4 + 2x^2 + 1) + x^2 + 1 = 2x^4 + 5x^2 + 3$$

c) To work out ff(x) we apply the 'f' rule first then the 'f' rule again:

$$ff(x) = f(x^2 + 1) = (x^2 + 1)^2 + 1$$
$$= x^4 + 2x^2 + 2$$

d) To work out gg(x) we apply the 'g' rule first then the 'g' rule again:

$$g(g(x)) = g(2x^2 + x) = 2(2x^2 + x)^2 + (2x^2 + x)$$
$$= 2(4x^4 + 4x^3 + x^2) + (2x^2 + x) = 8x^4 + 8x^3 + 4x^2 + x$$

Example

Functions f and g are defined by:

$f(x) = \sqrt{x - 2}$ $\{x: x \geq 2\}$ and $g(x) = x^2 + 4$ $\{x: x \geq 0\}$.

a) Find an expression for gf(x).

b) State the domain of gf(x).

c) State the range of gf(x).

Answer

a) To work out gf(x) we apply the 'f' rule first then the 'g rule'

Step 1: State the 'f' rule which is $f(x) = \sqrt{x - 2}$

Step 2: Apply the 'g 'rule to the result in step 1 to give $g(f(x))$.

The 'g' rule states that you square whatever is the object (now $\sqrt{x - 2}$) and add on 4:

$$gf(x) = g(\sqrt{x - 2})$$
$$= (\sqrt{x - 2})^2 + 4$$

Step 3: Simplify the algebra $\Rightarrow$ gf$(x) = (x - 2) + 4 = x + 2$

b) The domain of gf(x) is the domain of f since gf(x) means we apply the 'f' rule first : this becomes the new 'object' for us to apply the 'g' rule. For $f(x)$ to have real values since $f(x) = \sqrt{x - 2}$, then $x \geq 2$ so the domain is $\{x: x \geq 2\}$.

c) $f(x) = \sqrt{x - 2}$ and since $x \geq 2$ then $f(x) \geq 0$

Let $y = gf(x)$ so we need to find any restrictions on the values for y as this is then the range of gf(x).

$y = gf(x) = x + 2$ from (a) but if $x \geq 2$ then $y \geq 4$

Therefore the range of gf(x) is $\{y: y \geq 4\}$.

Inverse functions

A function has been defined as a mapping from a set of elements in the **domain** to a set of elements in the **range**. For each element in the domain there is exactly one element in the range. For example, if $f(x) = x^2$ for $x \in \Re$ then for this mapping all the real numbers are squared so:

-2 is mapped to 4

-1 is mapped to 1

2 is mapped to 4

3 is mapped to 9

2.4 is mapped to 5.76

and so on.

Suppose that we begin with an element in the range, for example f(x) = 4, and we want to find the corresponding value of x in the domain. There are two values of x for which f(x) = 4; these are $x = -2$ and $x = +2$

We can therefore conclude that f(x) = x^2 for $x \in \Re$ is a many-one function (or mapping).

This process of 'going backwards' from the range to the domain is called the 'inverse'. The process only leads to a unique answer in the domain if the original function is one-one.

f(x) = x^2 can give a one-one function but only if we restrict the domain to $\{x: x \in \Re^+\}$ which means the domain is just the positive real numbers. If we let the domain be all real numbers including negative numbers then we have seen above that the function would be many-one.

Using this restricted domain f(x) = x^2 is one-one and therefore to each x there is a unique y = f(x) = x^2 and to each y there is a unique $x = \sqrt{y}$.

Definition

For a one-one function f(x) each element in the domain set A is mapped onto exactly one element in the range set B. The **inverse function** is the mapping which maps elements in the range set B back to elements in the domain set A.

The notation for the inverse function is $f^{-1}(x)$.

Figure 1.14 below shows the relationship between a function and its inverse.

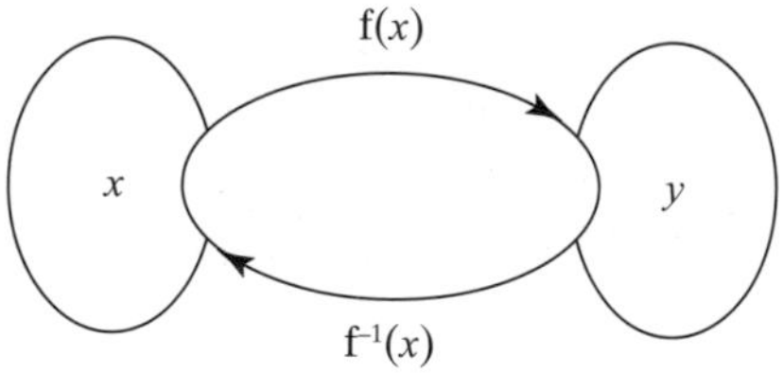

Method notes

If $y = x^2$ then taking the square root of both sides of this equation gives $\sqrt{y} = x$.

Essential notes

You must learn the inverse function notation **$f^{-1}(x)$** and write it correctly. It is **not** to be written as **$(f(x))^{-1}$**.

Fig. 1.14
Relationship between f(x) and f^{-1} (x).

Example
Find the inverse function of the function

$$f(x) = \frac{2}{x - 5} \ \{x \in \Re, \ x \neq 5\}$$

Answer
Step 1: Let y = f(x) = $\dfrac{2}{x - 5}$

Step 2: Rewrite as $(x-5)\,y=2$

Step 3: Divide both sides by y: $x-5=\dfrac{2}{y}$

Step 4: Make x the subject of the equation: $x=\dfrac{2}{y}+5$

Step 5: Interchange x and y in the result of step 4

$$y=\dfrac{2}{x}+5$$

Step 6: Write y as the inverse function:

$$f^{-1}(x)=\dfrac{2}{x}+5=\dfrac{2+5x}{x}$$

Essential notes

The subject of an equation is the variable written in terms of the other variables of the equation.

If $x=y^2+7$ then x is the subject.

If $y=x^3+2x$ then y is the subject.

Figure 1.15 shows the graphs of the given function $y=f(x)=\dfrac{2}{x-5}$ (blue) and its inverse function $y=f^{-1}(x)=\dfrac{2+5x}{x}$ (red).

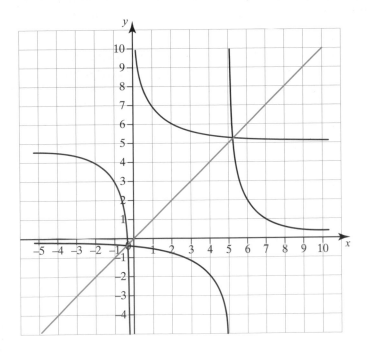

Fig. 1.15

Graphs of $y=f(x)=\dfrac{2}{x-5}$ (blue) and $y=f^{-1}(x)=\dfrac{2+5x}{x}$ (red).

Figure 1.15 also shows the line $y=x$ as the green line. From this diagram you can see that the graph of $f^{-1}(x)$ is a reflection of the graph of $f(x)$ in this line $y=x$.

Essential notes

The graphs of $\log(x)$ and 10^x were explained in Core 2. One graph was the reflection of the other graph in the line $y=x$. This shows that $\log(x)$ is the inverse function of 10^x.

Example

a) Find the inverse function, $f^{-1}(x)$, for the function
$$f(x) = \sqrt{5 - x} \; \{x \in \Re, x \leq 5\}$$

b) Simplify $f\,f^{-1}(x)$.

Answer

a) **Step 1**: Let $y = f(x) = \sqrt{5 - x}$

Step 2: Rewrite the equation in step 1 by squaring both sides:
$$y^2 = 5 - x$$

Step 3: Make x the subject of the equation:
$$x = 5 - y^2$$

Step 4: Interchange x and y in the equation of step 3
$$y = 5 - x^2$$

Step 5: From step 4 write y as the inverse function:
$$f^{-1}(x) = 5 - x^2$$

b) Using the result in (a) $f\,f^{-1}(x)$ means we apply the f^{-1} rule first to x so:
$$f^{-1}(x) = 5 - x^2$$

Then we apply the f rule to $(5 - x^2) \Rightarrow f(5 - x^2)$. The f rule is $\sqrt{5 - object}$ where the 'object' is now $(5 - x^2)$ so:
$$f\,f^{-1}(x) = \sqrt{5 - (5 - x^2)} = \sqrt{x^2} = x$$

Stop and think 2

If $f(x) = \sqrt{5 - x} \; \{x \in \Re, x \leq 5\}$ is $f\,f^{-1}(x) = f^{-1}\,f(x)$?

The modulus function

You will be familiar with the modulus sign $|\,|$ from GCSE. The **modulus** of a number is its **absolute** value i.e. its numerical value, ignoring its sign. For example

$$|1.7| = 1.7 \quad |{-3.1}| = 3.1$$

Algebraic equations and inequalities can also involve the use of the modulus sign. For example, $|x| = 4$ which we read as the modulus of x is 4 means that $x = -4$ or $x = 4$

$|x| < 4$ which we read as the modulus of x is less than 4 means -4 and $+4$ are the boundary values so $-4 < x < 4$

A **modulus function** is defined as $y = |f(x)|$. This means that y will always be positive whatever the sign of $f(x)$. When $f(x) \geq 0$, $|f(x)| = f(x)$. When $f(x) < 0$, $|f(x)| = f(x)$.

Essential notes

Absolute value means the positive real value.

Figure 1.16 shows the graphs of the linear function $y = x$ and the modulus function $y = |x|$.

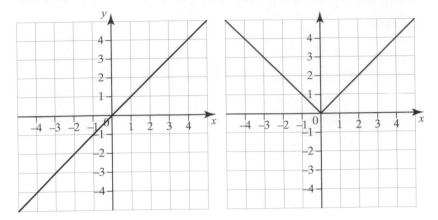

Essential notes

When drawing the graph of a modulus function $y = |f(x)|$, the graph will never go below the x-axis since $y = |f(x)|$ is always positive.

Fig. 1.16
The graphs of the linear function $y = x$ (blue) and the modulus function $y = |x|$ (red).

In drawing the graph of the modulus function $y = |x|$ from the linear function $y = x$, the part of the blue line below the x-axis is reflected in the x-axis. Another way of remembering this is that once a modulus graph reaches the x-axis it must be reflected in the x-axis and therefore be above the x-axis from that point. y can never be negative so the graph cannot be below the x-axis.

Example
Sketch the graph of $y = |x^2 - 7x + 6|$.

Answer

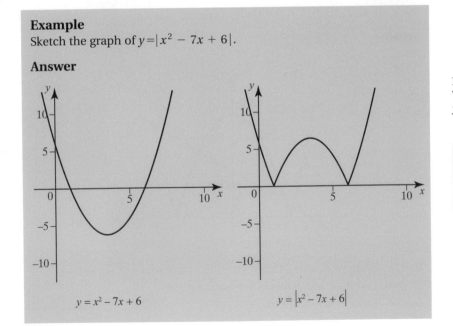

$y = x^2 - 7x + 6$

$y = |x^2 - 7x + 6|$

Fig. 1.17
The graphs of $y = x^2 - 7x + 6$ and $y = |x^2 - 7x + 6|$.

Method notes

Step 1: Sketch the graph of $y = x^2 - 7x + 6$ by finding the x-intercepts, ie where $y = 0$. So $x^2 - 7x + 6 = (x - 6)(x - 1) = 0$ therefore $x = 6$, $x = 1$.

Step 2: Reflect any part which is below the x-axis in the sketch graph above the x-axis in the modulus graph.

Fig. 1.18
The graphs of $y = x^2 - x - 6$ for $x \geq 0$ and $y = |x|^2 - |x| - 6$.

Method notes

Step 1: Sketch the graph of $y = x^2 - x - 6$ for $x \geq 0$ by finding where the graph crosses the x-axis i.e. where $y = 0$, so $x^2 - x - 6 = (x - 3)(x + 2) = 0$ therefore $x = 3$, $x = -2$

We can ignore $x = -2$ for this question since $x \geq 0$ So the graph crosses the x-axis at $(3, 0)$.

Step 2: For the modulus graph reflect the graph from step 1 in the y-axis since $|x|^2 - |x| - 6$ means x can have positive or negative values. In other words the graph will be symmetrical about the y-axis.

Fig. 1.19
The graphs of of $y = |x^2 - 9|$ and $y = x + 4$

Method notes

a) **Step 1**: Sketch $y = x^2 - 9$ Reflect any part of this graph which is below the x-axis in the x-axis, to give the graph of $y = |x^2 - 9|$ as shown in red in Fig. 1.19

   Step 2: Sketch $y = x + 4$ as shown in blue in Fig. 1.19

b) The solution of the equation $|x^2 - 9| = x + 4$ is where the two graphs intersect. There are four points of intersection of these two graphs so there are four solutions of the equation.

Example
Sketch the graph of $y = |x|^2 - |x| - 6$

Answer

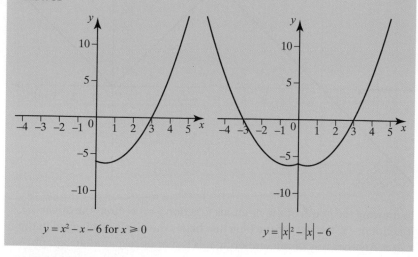

$y = x^2 - x - 6$ for $x \geq 0$

$y = |x|^2 - |x| - 6$

Example
On the same diagram, sketch the graphs of $y = |x^2 - 9|$ and $y = x + 4$

Solve the equation $|x^2 - 9| = x + 4$ correct to 2 d.p.

Answer
a)

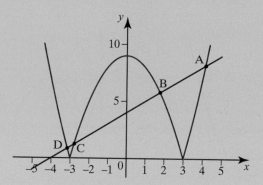

b) The solutions at A and D are where the non-reflected part of the (red) graph $y = x^2 - 9$ meets the line $y = x + 4$

So we solve the equation

$x^2 - 9 = x + 4$

$\Rightarrow \quad x^2 - x - 13 = 0$

$\Rightarrow \quad x = \dfrac{1 \pm \sqrt{(-1)^2 - 4 \times 1 \times (-13)}}{2}$

$\Rightarrow \quad x = \dfrac{1 \pm \sqrt{53}}{2}$

$\Rightarrow \quad x = 4.14 \text{ (point A) or } x = -3.14 \text{ (point D)}$

The solutions at B and C are where the reflected part (red) of the original graph of the function $y = x^2 - 9$ meets the line $y = x + 4$. This reflected part is therefore the graph of the function $y = -(x^2 - 9)$.

So we solve the equation

$-(x^2 - 9) = x + 4$

$\Rightarrow \quad x^2 + x - 5 = 0$

$\Rightarrow \quad x = \dfrac{-1 \pm \sqrt{1^2 - 4 \times 1 \times (-5)}}{2}$

$\Rightarrow \quad x = \dfrac{-1 \pm \sqrt{21}}{2}$

$\Rightarrow \quad x = 1.79 \text{ (point B) or } x = -2.79 \text{ (point C)}$

Transformations of graphs

Many graphs of functions can be described in terms of simple transformations of the graphs of standard functions such as polynomials and reciprocal functions.

In Core 1 you met the transformations of graphs which are summarised below.

Translation of a graph

$f(x + a)$ is a **horizontal** translation of $-a$ units

$f(x - a)$ is a **horizontal** translation of a units

$f(x) + a$ is a **vertical** translation of $+a$ units

$f(x) - a$ is a **vertical** translation of $-a$ units

Stretch of a graph

$af(x)$ is a **vertical** stretch factor a

$f(ax)$ is a **horizontal** stretch factor $\dfrac{1}{a}$

In this chapter combinations of two or more of these transformations are applied to curves.

Example

Use two transformations of the graph of $y = x^2$ to sketch the graph of $y = (x+2)^2 + 1$

Answer

Step 1: Sketch the graph of $y = x^2$ which is called the 'basic' function.

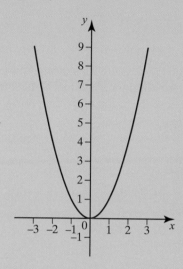

Fig. 1.20
The graph of $y = x^2$.

Step 2: Replace 'x' in the basic function by $(x+2)$ so the graph in step 1 is translated 2 units to the left.

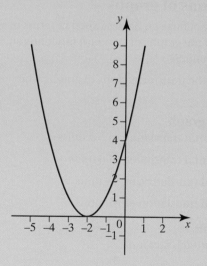

Method notes

It is often helpful to think of the 'pLus' 2 in $(x+2)^2$ as the graph 'moving to the Left' 2 units.

Fig. 1.21
The graph of $y = (x+2)^2$.

Step 3: Add 1 to $y = (x+2)^2$ so the graph in step 2 has a vertical translation of $+1$

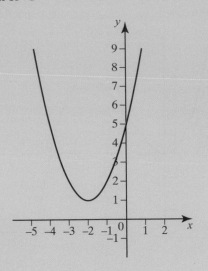

Fig. 1.22
The graph of $y = (x+2)^2 + 1$

Example
Sketch the graph of $y = |3x| - 1$

Answer
Use the basic equation $y = |x|$.

Step 1: $y = |3x|$ is a horizontal stretch of the basic function graph , scale factor $\dfrac{1}{3}$

Step 2: $y = |3x| - 1$ is then a vertical translation of -1 of the graph in step 1 of one unit downwards.

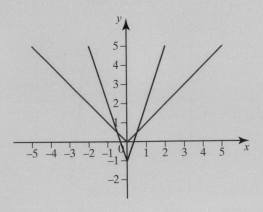

Method notes

To sketch $y = |x|$ draw the graph of $y = x$ and reflect in the x-axis, any part of this graph which is below the x-axis as shown in red in Fig 1.23

For $y = |3x|$ a horizontal stretch of scale factor $\dfrac{1}{3}$ means that the basic function graph is 'squeezed' at each point i.e. the x-values are multiplied by $\dfrac{1}{3}$ so the point $(1, 1)$ becomes the point $\left(\dfrac{1}{3}, 1\right)$. The graph of $y = |3x| - 1$ is shown in blue in Fig. 1.23

Fig. 1.23
Graphs of $y = |x|$ (red) and $y = |3x| - 1$ (blue).

Example

The diagram shows the graph of a function $y = f(x)$. The points A, B and C have coordinates (0, 0), (2, 4) and (3, 0) respectively.

Fig. 1.24
The graph of a function $y = f(x)$.

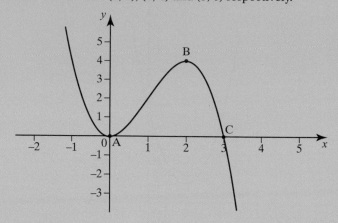

Method notes

In this example it does not matter that we do not know the function $f(x)$ explicitly because we are given the shape of the graph as shown in Fig. 1.24

Sketch the graph of:

a) $y = 2f(x) + 1$

b) $y = f(2x) + 1$

c) $y = f(x + 2) - 3$

d) $y = | f(x) - 2 |$

In each part state the coordinates of the points A, B and C after the transformation.

Answer

a)

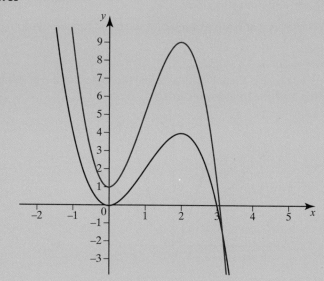

Fig. 1.25
Graphs of $y = f(x)$ (blue) and $y = 2f(x) + 1$ (red).

Method notes

a) To deal with any 'changes' to the basic function we start with the effect of those which are written 'nearest' to $f(x)$.

Start with the effect of $2f(x)$ which is a vertical stretch factor of 2, followed by a vertical translation of +1

The x-values at A, B and C are unchanged.

The y-values of A, B and C are each multiplied by 2 and then have 1 added.

Point A moves from (0, 0) to $(0, 0 + 1) = (0, 1)$

Point B moves from (2, 4) to $(2, 2 \times 4 + 1) = (2, 9)$

Point C moves from (3, 0) to $(3, 2 \times 0 + 1) = (3, 1)$

b)

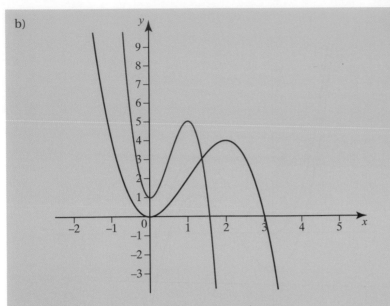

Point A moves from $(0, 0)$ to $\left(0 \times \dfrac{1}{2}, 0 + 1 \right) = (0, 1)$

Point B moves from $(2, 4)$ to $\left(2 \times \dfrac{1}{2}, 4 + 1 \right) = (1, 5)$

Point C moves from $(3, 0)$ to $\left(3 \times \dfrac{1}{2}, 0 + 1 \right) = \left(1\dfrac{1}{2}, 1 \right)$

c)

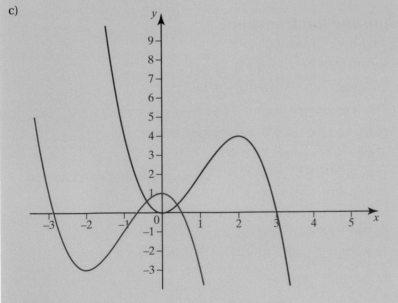

Point A moves from $(0, 0)$ to $(0 - 2, 0 - 3) = (-2, -3)$

Point B moves from $(2, 4)$ to $(2 - 2, 4 - 3) = (0, 1)$

Point C moves from $(3, 0)$ to $(3 - 2, 0 - 3) = (1, -3)$

Fig. 1.26
Graphs of $y = f(x)$ (blue) and
$y = f(2x) + 1$ (red).

Method notes

b) Start with the effect
of $f(2x)$ which is a
horizontal stretch factor
of $\dfrac{1}{2}$, followed by a vertical
translation of $+1$

The x-values at A, B and C
are each multiplied by $\dfrac{1}{2}$

The y-values at A, B and C
are each increased by 1

c) Start with the effect of
$f(x + 2)$ which moves the
curve 2 units 'to the left'
followed by a vertical
translation of -3 units.
The x-values at A, B and C
each have 2 subtracted and
the y-values each have 3
subtracted.

Fig. 1.27
Graphs of $y = f(x)$ (blue) and
$y = f(x + 2) - 3$ (red).

Fig. 1.28
Graphs of $y = f(x)$ (blue), $y = f(x) - 2$ (red) and $y = |f(x) - 2|$ (green).

Method notes

d) Start with the effect of $f(x) - 2$ which means the curve $y = f(x)$ is translated vertically by -2 units to give $y = f(x) - 2$ (shown in Fig. 1.28 by the red curve) then deal with the effect of the modulus which means any part of the red curve below the x-axis is reflected in the x-axis, as shown by the green curve.

The x-values at A, B and C are unchanged.

The y-values at A, B and C become $|y$ value at point $-2|$.

d)

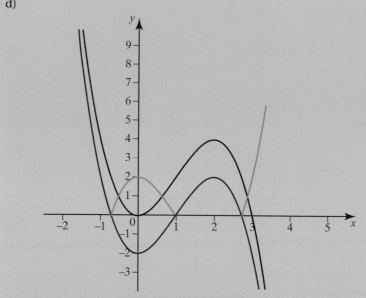

Point A moves from $(0, 0)$ to $(0, |0 - 2|) = (0, 2)$

Point B moves from $(2, 4)$ to $(2, |4 - 2|) = (2, 2)$

Point C moves from $(3, 0)$ to $(3, |0 - 2|) = (3, 2)$

Stop and think answers

1 Let $f(x) = x + 3$ and $g(x) = x - 3$

then $f(g(x)) = f(x - 3) = (x - 3) + 3 = x$

and $g(f(x)) = g(x + 3) = (x + 3) - 3 = x$

This is an example where it is true that $f(g(x)) = g(f(x))$

In this case $f(x)$ and $g(x)$ were inverse functions.

2 If $f(x) = \sqrt{(5 - x)}$ and $x \leq 5$

Then let $y = \sqrt{(5 - x)} \Rightarrow y^2 = (5 - x)$ therefore making x the subject of this equation gives $x = 5 - y^2$

Interchanging x and $y \Rightarrow y = 5 - x^2$ the writing y as the inverse function $f^{-1}(x) = 5 - x^2$.

To evaluate $f(f^{-1}(x)) = f(5 - x^2) = \sqrt{5 - (5 - x^2)} = \sqrt{x^2} = x$.

To evaluate $f^{-1}(f(x)) = f^{-1}\left(\sqrt{5 - x}\right) = 5 - \left(\sqrt{5 - x}\right)^2$
$= 5 - (5 - x) = x$.

Trigonometric functions

You are familiar with the definitions and graphs of the basic trigonometric functions sin x, cos x and tan x.

The functions sin x and cos x have the set of real numbers as their domain and their range is $\{y: -1 \leq y \leq 1\}$.

You will recall that the function tan x is not valid for the following values

$$x = ..., -270°, -90°, 270°, ... \text{ or in radians, } x = ..., -\frac{3\pi}{2}, -\frac{\pi}{2}, \frac{\pi}{2}, \frac{3\pi}{2}, ...$$

There must be vertical asymptotes at these x-values on the graph of $y = \tan x$.

The values of x for which tan x is not valid can be generated by the formula $x = 90° \pm 180°n$ for $n \in Z$ where Z means the set of integers.

So if $n = 1$, $x = 90° \pm 180°$

$$= 270°, -90°$$

Taking $n = 2, 3, ...$ gives the rest of the x-values for which tan x is not valid.

If x is in radians $\left(\dfrac{\pi}{2} = 90°\right)$ the x-values for which tan x is not valid are generated by the formula $x = \dfrac{\pi}{2} \pm \pi n$ for $n \in Z$.

Graphs of the trigonometric functions show that the function values are repeated at regular intervals and so they are called **periodic functions**.

- cos x repeats itself every 360° so we say the period is 360° (or 2π radians) as shown in Figure 2.1.

- sin x repeats itself every 360° so we say the period is 360° (or 2π radians) as shown in Figure 2.2.

- tan x repeats itself every 180° so we say the period is 180° or π radians as shown in Figure 2.3.

Fig. 2.1
Graph of $y = \cos x$.

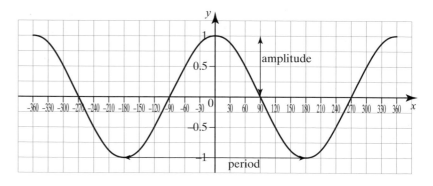

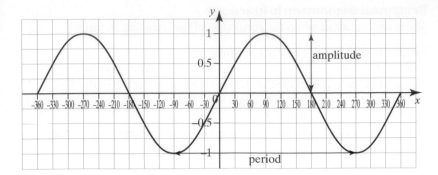

Fig. 2.2
Graph of $y = \sin x$.

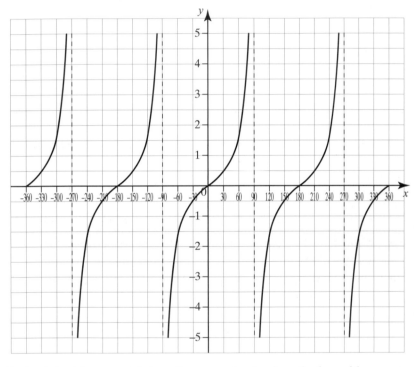

Fig. 2.3
Graph of $y = \tan x$.

Each of the trigonometric functions can also be described as odd or even. Even and odd functions were explained in Core 1:

- If $f(x) = f(-x)$ then the function is 'even'.

- If $f(x) = \cos x$ then $\cos x = \cos(-x)$ for all values of x.

- For example, $\cos 60° = \cos(-60°) = 0.5$ therefore $\cos x$ is an 'even' function.

- If $f(x) = -f(-x)$ then the function is 'odd'.

- If $f(x) = \sin x$ then $\sin x = -\sin(-x)$ for all values of x.

- For example, $\sin 30° = -\sin(-30°) = 0.5$ therefore $\sin x$ is an 'odd' function.

- If $f(x) = \tan x$ then $\tan x = -\tan(-x)$ for all values of x.

- For example, $\tan 45° = -\tan(-45°) = 1$ therefore $\tan x$ is an 'odd' function.

Reciprocal trigonometric functions

There are occasions when the reciprocal of the basic trigonometric functions, such as $\dfrac{1}{\cos x}$, occur in problems.

An example would be in mechanics when considering the path of an object which is projected at an angle θ to the horizontal. Such a path would be a parabola and its equation is known to be $y = x \tan \theta - \dfrac{gx^2}{2u^2 \cos^2 \theta}$.

This equation includes the expression $\dfrac{1}{\cos^2 \theta}$ which is the reciprocal of $(\cos \theta)^2$.

The three reciprocals of the basic trigonometric functions $\sin x$, $\cos x$ and $\tan x$ lead to three new trigonometric functions:

- $\sec x = \dfrac{1}{\cos x}$ called the **secant**

- $\operatorname{cosec} x = \dfrac{1}{\sin x}$ called the **cosecant**

- $\cot x = \dfrac{1}{\tan x}$ called the **cotangent**.

Each of the angles in the following example is one of the 'common' angles and you are expected to know the three basic trigonometric ratios of all common angles. You are also expected to know whether a trigonometric ratio is positive or negative in each of the four quadrants as explained in Core 2. These are shown in the diagram below.

Essential notes

These three new functions may have their own button on your calculator. If not, to find $\sec 30°$ you can work this out by using $\dfrac{1}{\cos 30°}$.

Fig. 2.4

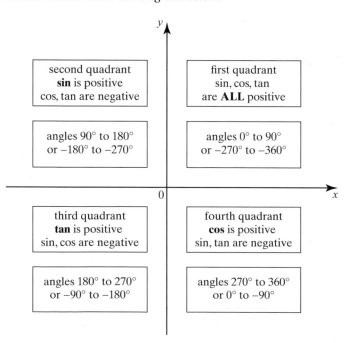

Example

Without using a calculator find the **exact** values of

a) cosec 30° b) 5 sec² 135° c) cot (−60°).

Answer

a) $\operatorname{cosec} 30° = \dfrac{1}{\sin 30°} = \dfrac{1}{\frac{1}{2}} = 2$

b) **Step 1:** $\sec^2 135° = \dfrac{1}{(\cos 135)^2} = \dfrac{1}{\frac{-1}{\sqrt{2}}} \times \dfrac{1}{\frac{-1}{\sqrt{2}}} = \dfrac{1}{\frac{1}{2}} = 2$

Step 2: $5 \sec^2 135° = 5 \times \dfrac{1}{\cos^2 135°} = 5 \times \dfrac{1}{\frac{1}{2}} = 10$

c) $\cot(−60°) = \dfrac{1}{\tan(−60°)} = \dfrac{1}{−\sqrt{3}} = −\dfrac{\sqrt{3}}{3}$

Example

Find the values of

a) $\sec \dfrac{\pi}{3}$ b) $\operatorname{cosec} \dfrac{2\pi}{3}$ c) $\sec \dfrac{\pi}{2}$

Answer

Check that your calculator is in radians mode.

a) $\sec \dfrac{\pi}{3} = \dfrac{1}{\cos \frac{\pi}{3}} = \dfrac{1}{0.5} = 2$

b) $\operatorname{cosec} \dfrac{2\pi}{3} = \dfrac{1}{\sin \frac{2\pi}{3}} = \dfrac{1}{\frac{\sqrt{3}}{2}} = \dfrac{2}{\sqrt{3}}$

or $\operatorname{cosec} \dfrac{2\pi}{3} = \dfrac{1}{\sin \frac{2\pi}{3}} = 1.1547$ (using a calculator)

c) $\sec \dfrac{\pi}{2} = \dfrac{1}{\cos \frac{\pi}{2}} = \dfrac{1}{0}$ which is not defined.

The answers here can be given as surds or decimals since the question did not ask for 'exact' values.

In part c) of the last example you saw that sec x is not defined for $x = \dfrac{\pi}{2}$ because $\sec \dfrac{\pi}{2} = \dfrac{1}{\cos \frac{\pi}{2}} = \dfrac{1}{0}$ which approaches ∞. We therefore say that the function sec x is undefined for $x = \dfrac{\pi}{2}$. We can develop this further to show that the reciprocal trigonometric functions are not defined for all x-values.

Essential notes

Exact means that you should state the answer in terms of surds, not decimals. Surds were explained in Core 1. Trigonometric ratios of common angles and positive or negative angles were explained in Core 2.

Method notes

a) $\sin 30° = \dfrac{1}{2} = 0.5$

b) $\cos 135° = −\cos 45° = −\dfrac{\sqrt{2}}{2}$

$= −\dfrac{1}{\sqrt{2}}$

$\cos^2 135° = \left(−\dfrac{1}{\sqrt{2}}\right)^2 = \dfrac{1}{2}$

c) $\tan(−60°) = −\tan 60°$

$= −\sqrt{3}$

Essential notes

When angles are given in terms of π this means you must use your calculator in radians mode.

You must recognise common angles in radians: $\dfrac{\pi^c}{3} = 60°$,

$\dfrac{\pi^c}{2} = 90°$, $\pi^c = 180°$.

- $\sec x = \dfrac{1}{\cos x}$ is not defined where $\cos x = 0$ i.e. for $\{x: x = 90° \pm 180n°\}$

  or, in radians, $\{x: x = \dfrac{\pi}{2} \pm n\pi\}$

- $\operatorname{cosec} x = \dfrac{1}{\sin x}$ is not defined where $\sin x = 0$ i.e. for $\{x: x = \pm 180n°\}$ or,

  in radians, $\{x: x = \pm n\pi\}$

- $\cot x = \dfrac{1}{\tan x}$ is not defined where $\tan x = 0$ i.e. for $\{x: x = \pm 180n°\}$ or, in

  radians, $\{x: x = \pm n\pi\}$

Equations involving reciprocal trigonometric functions

Method notes

Use the definitions to change the equations from the reciprocal trigonometric functions to the basic trigonometric functions. Then solve in the usual way.

You can use a sketch graph to identify the secondary solutions as shown below, or the CAST diagram.

a)

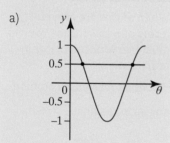

Fig. 2.5

b)

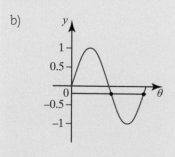

Fig. 2.6

Example

Find values of θ which satisfy the following equations such that $0° \le \theta \le 360°$:

a) $\sec \theta = 2$

b) $\operatorname{cosec} \theta = -5$

c) $1 + \cot^2 \dfrac{1}{2}\theta = 5$

Answer

a) $\sec \theta = 2$

$$\Rightarrow \frac{1}{\cos\theta} = 2$$

$$\Rightarrow \cos\theta = 0.5$$

$$\Rightarrow \quad \theta = 60°$$

(this the primary solution from a calculator)

but $0° \le \theta \le 360°$ and cosine is also positive in the fourth quadrant therefore

$$\theta = 360° - 60°$$

$$= 300°$$

which gives the secondary solution.

b) $\operatorname{cosec} \theta = -5$

$$\Rightarrow \frac{1}{\sin\theta} = -5$$

$$\Rightarrow \sin\theta = -0.2$$

$$\Rightarrow \quad \theta = -11.5°$$

This primary solution from the calculator is in the fourth quadrant but $0° \le \theta \le 360°$ and sine is also negative in the third quadrant

$\Rightarrow \theta = 180° + 11.5 = 191.5°$ or

$\quad \theta = 360° - 11.5°$

$\qquad = 348.5°$

which gives the secondary solution.

c) **Step 1:** In this part of the question you must solve for θ although the angle in the original equation is $\frac{1}{2}\theta$.

Step 2: To solve $1 + \cot^2 \frac{1}{2}\theta = 5$ simplify the equation to $\cot^2 \frac{1}{2}\theta = 4$

Step 3: Take the square root of both sides of the equation so $\cot \frac{1}{2}\theta = \pm 2$

Step 4: $\cot \frac{1}{2}\theta = \pm 2$

$\Rightarrow \dfrac{1}{\tan \dfrac{\theta}{2}} = \pm 2$

$\Rightarrow \tan \frac{1}{2}\theta = \pm \frac{1}{2}$

If $\tan \frac{1}{2}\theta = +0.5$

$\Rightarrow \quad \frac{1}{2}\theta = 26.56°$

$\Rightarrow \qquad \theta = 53.1°$

which is a primary solution but $0° \le \theta \le 360°$ and the tangent is also positive in the third quadrant so:

$\frac{1}{2}\theta = 180° + 26.56°$

$\Rightarrow \frac{1}{2}\theta = 206.56°$

$\Rightarrow \quad \theta = 2 \times 206.56° = 412.56°$

which is outside the given range $0° \le \theta \le 360°$

If $\tan \frac{1}{2}\theta = -0.5$

$\Rightarrow \quad \frac{1}{2}\theta = -26.56°$

which is a primary solution but $0° \le \theta \le 360°$ and tangent is also negative in the second quadrant

$\Rightarrow \frac{1}{2}\theta = 180° - 26.56° = 153.44°$

$\Rightarrow \quad \theta = 306.8°$

So the answers are $\theta = 53.1°$ and $306.8°$

c) Here you only need to sketch the tangent graph up to 180°. This helps to locate solutions for $\frac{1}{2}\theta$.

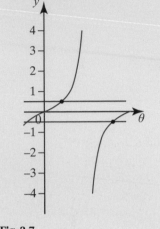

Fig. 2.7

Graphs of the reciprocal trigonometric functions

The definitions of sec x, cosec x and cot x are derived from the trigonometric functions cos x, sin x and tan x respectively. We use the graphs of these trigonometric functions to sketch the graphs of the **reciprocal trigonometric functions**.

To draw the graph of sec x considering its values of sec x between 0° and 180° helps us find where in the x-y plane the graph lies.

$$\sec 0° = \frac{1}{\cos 0°} = 1$$

$$\sec 30° = \frac{1}{\cos 30°} = 1.155$$

$$\sec 45° = \frac{1}{\cos 45°} = 1.414$$

$$\sec 60° = \frac{1}{\cos 60°} = 2$$

$$\sec 90° = \frac{1}{\cos 90°} = \frac{1}{0} \text{ (undefined)}$$

$$\sec 120° = \frac{1}{\cos 120°} = -2$$

$$\sec 135° = \frac{1}{\cos 135°} = -1.414$$

$$\sec 150° = \frac{1}{\cos 150°} = -1.155$$

$$\sec 180° = \frac{1}{\cos 180°} = -1$$

From these results we can conclude the following properties of sec x:

- from the cosine graph $-1 \le \cos x \le 1$ for all values of x therefore as $\sec x = \dfrac{1}{\cos x}$ you can deduce that $\sec x \ge 1$ and $\sec x \le -1$ for all values of x.

- $\sec 90° = \dfrac{1}{\cos 90°} = \dfrac{1}{0}$ which is undefined therefore there is a vertical asymptote at 90° and all other values of x where $\cos x = 0$

- substituting values of n into the formula $x = \pm 360n°$ for $n \in Z$, shows that $\sec x = 1$ at the points where $x = 0, \pm 360°$

- substituting values of n into the formula $x = 180° \pm 360n°$ for $n \in Z$, shows that $\sec x = -1$ at the points where $x = \pm 180°$

- The graph of $y = \sec x$ is periodic with period 360° because the graph of $y = \cos x$ is periodic with period 360°.

- $y = \sec x$ is an even function because $y = \cos x$ is an even function.

Essential notes

Z is the set of all integers.

Periodic functions were explained earlier in this chapter.

Even and odd functions were explained earlier in this chapter.

An asymptote is a line along which a curve approaches infinity.

Figure 2.8 shows these properties on the graph of $y = \sec x$ for the domain $\{x: -360° \leq x \leq 360°\}$. Note the asymptotes at $-270°$, $-90°$, $90°$, $270°$.

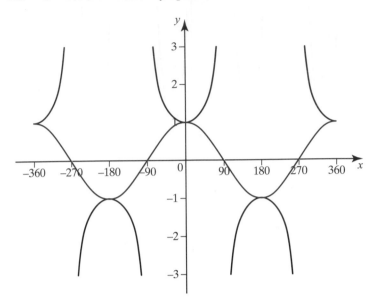

Fig. 2.8
The graphs of $y = \cos x$ (red) and $y = \sec x$ (blue).

The graphs of $y = \operatorname{cosec} x$ and $y = \cot x$ show similar features and are shown in figures 2.9 and 2.10.

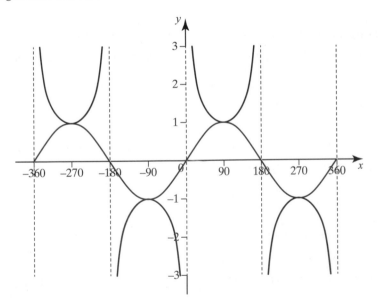

Fig. 2.9
The graphs of $y = \sin x$ (red) and $y = \operatorname{cosec} x$ (blue).

Essential notes

The graph of $y = \operatorname{cosec} x$ (blue) has vertical asymptotes at $x = 180° \pm 360n°$ for $n \in Z$.

$\operatorname{cosec} x = 1$ when $x = 90° \pm 360n°$ for $n \in Z$.

$\operatorname{cosec} x = -1$ when $x = -90° \pm 360n°$ for $n \in Z$.

The graph of $y = \operatorname{cosec} x$ is periodic with period 360°.

$y = \operatorname{cosec} x$ is an odd function.

Fig. 2.10
The graphs of $y = \tan x$ (red)
and $y = \cot x$ (blue).

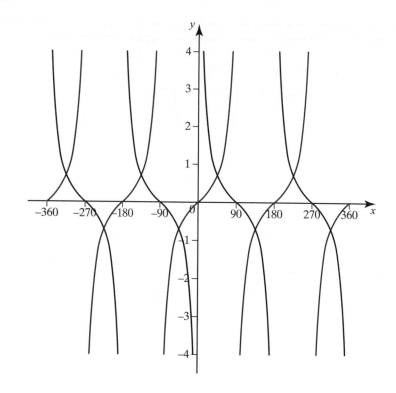

Essential notes

The graph of $y = \cot x$ (blue)
has vertical asymptotes at
$x = \pm 180n°$ for $n \in Z$.

At the values of x where
$y = \cot x$ crosses the x-axis,
$y = \tan x$ has asymptotes.

The graph of $y = \cot x$ is
periodic with period 180°.

$y = \cot x$ is an odd function.

Example

a) Complete this table to show the signs of the reciprocal trigonometric
   functions in the given domains.

	$0 < x < \dfrac{\pi}{2}$	$\dfrac{\pi}{2} < x < \pi$	$\pi < x < \dfrac{3\pi}{2}$	$\dfrac{3\pi}{2} < x < 2\pi$
sec x	+			
cosec x		+		
cot x			+	

b) State values of x in the interval $0 < x < 2\pi$ for which the following are
   not defined.

   i) sec x ii) cosec x iii) cot x

Answer

a) Using the graphs from the last section the table is completed as follows:

	$0 < x < \dfrac{\pi}{2}$	$\dfrac{\pi}{2} < x < \pi$	$\pi < x < \dfrac{3\pi}{2}$	$\dfrac{3\pi}{2} < x < 2\pi$
sec x	+	−	−	+
cosec x	+	+	−	−
cot x	+	−	+	−

Method notes

The graphs will help to
answer this example.

Remember that $360° \equiv 2\pi$.

b) The given interval is $0 < x < 2\pi$ so we must give the answer in radians. We can conclude the following as there are asymptotes where the graphs are undefined:

  i) sec x is undefined for $x = \dfrac{\pi}{2}$ and $x = \dfrac{3\pi}{2}$

  ii) cosec x is undefined for $x = \pi$

  iii) cot x is undefined for $x = \pi$

Further trigonometric identities

An important trigonometric identity involves sin x and cos x and it is useful for the manipulation of trigonometric expressions. This identity is:

$\sin^2 x + \cos^2 x \equiv 1$ **for all values of** x

It was introduced in Core 2 and is called the Pythagorean Identity. We can derive two further useful identities from the Pythagorean Identity.

Step 1: Divide $\sin^2 x + \cos^2 x \equiv 1$ throughout by $\cos^2 x$.

$$\Rightarrow \frac{\sin^2 x}{\cos^2 x} + \frac{\cos^2 x}{\cos^2 x} \equiv \frac{1}{\cos^2 x}$$

Step 2: Use the identities $\dfrac{\sin x}{\cos x} \equiv \tan x$ and $\dfrac{1}{\cos x} \equiv \sec x$

$$\Rightarrow \tan^2 x + 1 \equiv \sec^2 x \text{ for all values of } x$$

The third identity is derived from the Pythagorean Identity as follows.

Step 1: Divide $\sin^2 x + \cos^2 x \equiv 1$ throughout by $\sin^2 x$

$$\Rightarrow \frac{\sin^2 x}{\sin^2 x} + \frac{\cos^2 x}{\sin^2 x} \equiv \frac{1}{\sin^2 x}$$

Step 2: Use the identities $\dfrac{\cos x}{\sin x} \equiv \cot x$ and $\dfrac{1}{\sin x} \equiv \csc x$

$$\Rightarrow 1 + \cot^2 x \equiv \csc^2 x \text{ for all values of } x$$

To summarise, the three identities are:

$\sin^2 x + \cos^2 x \equiv 1$

$1 + \tan^2 x \equiv \sec^2 x$

$1 + \cot^2 x \equiv \csc^2 x$

They are useful for simplifying trigonometric expressions in proofs and also for solving equations.

Simplifying trigonometric equations

Example

Simplify $\dfrac{1 - \sec^2 \theta}{1 - \csc^2 \theta}$.

Method notes

Remember $\cot x = \dfrac{1}{\tan x}$

$$\Rightarrow \qquad \cot x = \frac{\cos x}{\sin x}$$

Exam tips

You must learn the three identities. They are not in your formula booklet.

Continued on the next page

Answer

Simplify means that we must try to write this in non-fraction form.

Step 1: Rewrite the identity $1 + \tan^2 \theta \equiv \sec^2 \theta$

$$\Rightarrow 1 - \sec^2 \theta \equiv -\tan^2 \theta$$

Step 2: Rewrite the identity $1 + \cot^2 \theta \equiv \text{cosec}^2 \theta$

$$\Rightarrow 1 - \text{cosec}^2 \theta \equiv -\cot^2 \theta$$

Step 3: Use the rewrites to simplify the numerator and denominator of the original statement:

$$\frac{1 - \sec^2 \theta}{1 - \text{cosec}^2 \theta} \equiv \frac{-\tan^2 \theta}{-\cot^2 \theta} \equiv \tan^4 \theta$$

Method notes

$$\cot^2 \theta = \left(\frac{1}{\tan \theta}\right)^2$$

Proving identities

Example

Prove the following identities:

a) $\dfrac{1}{\tan \theta + \cot \theta} \equiv \sin \theta \cos \theta$

b) $(\sin \theta + \text{cosec } \theta)^2 \equiv \cot^2 \theta - \cos^2 \theta + 4$

Answer

a) **Step 1:** Start with the LHS of the identity and rewrite

$$\frac{1}{\tan \theta + \cot \theta} \equiv \frac{1}{\tan \theta + \dfrac{1}{\tan \theta}}$$

$$\equiv \frac{\tan \theta}{\tan^2 \theta + 1} \equiv \frac{\tan \theta}{\sec^2 \theta}$$

Step 2: From the definition of $\sec \theta$, $\sec^2 \theta \equiv \left(\dfrac{1}{\cos \theta}\right)^2$

therefore substituting this into step 1

$$\Rightarrow \frac{\tan \theta}{\sec^2 \theta} = \tan \theta \cos^2 \theta$$

Step 3: $\tan \theta \equiv \dfrac{\sin \theta}{\cos \theta}$ therefore from step 2

$$\tan \theta \cos^2 \theta \equiv \frac{\sin \theta}{\cos \theta} \cos^2 \theta \equiv \sin \theta \cos \theta$$

which is the RHS of the identity.

Therefore the identity is proved.

b) **Step 1:** Start with the LHS of the identity and rewrite:

$$(\sin \theta + \text{cosec } \theta)^2 \equiv \sin^2 \theta + 2 \sin \theta \text{ cosec } \theta + \text{cosec}^2 \theta$$

Step 2: Use the definition of cosec θ from earlier in the chapter to rewrite the middle term of the answer in step 1, so

$$2 \sin \theta \text{ cosec } \theta \equiv 2 \sin \theta \times \frac{1}{\sin \theta} \equiv 2$$

Method notes

To prove an identity you must start with 'one side' of the identity statement and from that derive the 'other side' of the identity statement.

LHS means 'left hand side'.

RHS means 'right hand side'.

Use the trigonometric identity $\sec^2 \theta = \tan^2 \theta + 1$ which was derived earlier in this chapter.

Step 3: Rewrite the answer of step 1 using the rewrite in step 2

so $(\sin\theta + \operatorname{cosec}\theta)^2 = \sin^2\theta + 2 + \operatorname{cosec}^2\theta$

Step 4: $\operatorname{cosec}^2\theta = (1 + \cot^2\theta)$ and $\sin^2\theta = (1 - \cos^2\theta)$ therefore from step 3

$\sin^2\theta + 2 + \operatorname{cosec}^2\theta \equiv (1 - \cos^2\theta) + 2 + (1 + \cot^2\theta)$

Step 5: Simplify to give $\cot^2\theta - \cos^2\theta + 4 = $ RHS of the identity.

Therefore the identity is proved.

Method notes

Use the identity $\operatorname{cosec}^2\theta = (1 + \cot^2\theta)$ which was derived earlier in this chapter.

Example

a) Prove that $\sec^4 x - \tan^4 x \equiv \sec^2 x + \tan^2 x$

b) Hence solve the equation $\sec^4 x - \tan^4 x = 3\tan x$, giving the answers in the range $-\pi \le x \le \pi$.

Answer

a) **Step 1:** Start with the LHS of the identity and rewrite

$\sec^4 x - \tan^4 x$ as a 'difference of two squares' because

$\sec^4 x = (\sec^2 x)^2$ and $\tan^4 x = (\tan^2 x)^2$

hence $\sec^4 x - \tan^4 x \equiv (\sec^2 x + \tan^2 x)(\sec^2 x - \tan^2 x)$

Step 2: Rewrite the identity $1 + \tan^2 x \equiv \sec^2 x$

$\Rightarrow \qquad \sec^2 x - \tan^2 x \equiv 1$

Step 3: Rewrite the result from step 1 by substituting the result from step 2

$\Rightarrow \sec^4 x - \tan^4 x \equiv (\sec^2 x + \tan^2 x)(\sec^2 x - \tan^2 x)$

$\equiv (\sec^2 x + \tan^2 x)$ \hfill (1)

$\Rightarrow \sec^4 x - \tan^4 x \equiv (\sec^2 x + \tan^2 x)$ which is the RHS of the identity.

Therefore the identity is proved.

b) The LHS of the equation $\sec^4 x - \tan^4 x = 3\tan x$

is the same as the identity which has been proved in a).

Step 1: Use the result from a) to rewrite the equation:

$\sec^4 x - \tan^4 x = 3\tan x$

$\Rightarrow (\sec^2 x + \tan^2 x) = 3\tan x$

Step 2: Use the identity $1 + \tan^2 x \equiv \sec^2 x$ to rewrite the result of step 1

$\Rightarrow \quad (\sec^2 x + \tan^2 x) = (1 + \tan^2 x) + \tan^2 x = 3\tan x$

$\Rightarrow \qquad 1 + 2\tan^2 x = 3\tan x$

$\Rightarrow 2\tan^2 x - 3\tan x + 1 = 0$

Method notes

The difference of two squares was covered in Core 1. The factorisation of

$c^2 - d^2 = (c + d)(c - d)$

In (a) we take

$c = \sec^2 x$ and $d = \tan^2 x$

to factorise $\sec^4 x - \tan^4 x$.

Method notes

$2\tan x = 1 \Rightarrow \tan x = \dfrac{1}{2}$

We also know that $\tan x = 1$

The graph of $y = \tan x$,

$y = 1$ and $y = \dfrac{1}{2}$ is helpful in seeing how many solutions there are to the original equation and where they occur.

▮☞ Continued on the next page

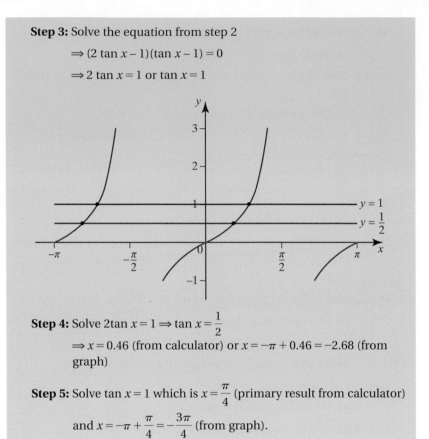

Step 3: Solve the equation from step 2

$$\Rightarrow (2\tan x - 1)(\tan x - 1) = 0$$

$$\Rightarrow 2\tan x = 1 \text{ or } \tan x = 1$$

Fig. 2.11

Graphs of $y = \tan x$, $y = 1$ and $y = \dfrac{1}{2}$

Step 4: Solve $2\tan x = 1 \Rightarrow \tan x = \dfrac{1}{2}$

$$\Rightarrow x = 0.46 \text{ (from calculator) or } x = -\pi + 0.46 = -2.68 \text{ (from graph)}$$

Step 5: Solve $\tan x = 1$ which is $x = \dfrac{\pi}{4}$ (primary result from calculator)

$$\text{and } x = -\pi + \frac{\pi}{4} = -\frac{3\pi}{4} \text{ (from graph)}.$$

Essential notes

The notation $\sin^{-1} x$, $\cos^{-1}$ and $\tan^{-1} x$ are also widely used for the inverse trigonometric functions.

This fits with the notation, $f^{-1}(x)$ used in chapter 1. Check which notation your calculator uses. You may use either notation.

This notation must not be confused with the reciprocal of the trigonometric functions

e.g. $(\sin x)^{-1} = \dfrac{1}{\sin x} = \csc x$.

Inverse trigonometric functions

In chapter 1 we defined the **inverse function** for a one-one function f(x) as:

'a mapping that maps elements in the range of f(x), back to elements in the domain of f(x)'.

The important feature of this definition is that the inverse function only exists for one-one functions. The trigonometric functions are many-one functions so their inverse functions will only exist if we restrict the domains. If we do this correctly, the functions are then one-one.

The notations used for the inverse functions of $\sin x$, $\cos x$ and $\tan x$ are respectively **arcsin x, arccos x and arctan x**.

Graph and properties of arcsin x

Figure 2.12 shows a graph of one cycle of the function

$f(x) = \sin x$. One cycle has domain $\{x: -\pi \leq x \leq \pi\}$.

Two values of x, $x = \dfrac{\pi}{6}$ and $x = \dfrac{5\pi}{6}$ in this domain give the same y-value in

the image because $\sin \dfrac{\pi}{6} = \dfrac{5\pi}{6} = 0.5$, so the function $f(x) = \sin x$ with this

domain is many-one. Inverse functions only exist for one-one functions so the inverse function cannot be defined for this domain $\{x: -\pi \leq x \leq \pi\}$.

Fig. 2.12

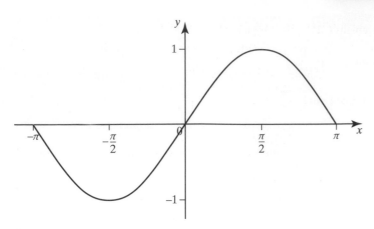

To define the inverse function arcsin x the domain of sin x is restricted to $\left\{x: -\dfrac{\pi}{2} \leq x \leq \dfrac{\pi}{2}\right\}$ which ensures that sin x is a one-one function. Figure 2.13 shows the graphs of f$(x) = \sin x$ and f$(x) = \arcsin x$.

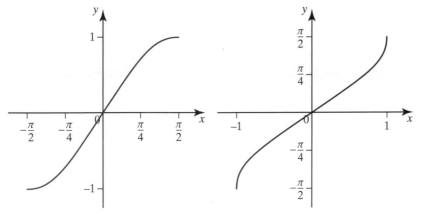

Fig. 2.13
Graph of $y = \sin x$ (red) and $y = \arcsin x$ (blue).

- the graph of $y = $ f$(x) = \arcsin x$ is a reflection of the graph of $y = $ f$(x) = \sin x$ in the line $y = x$.

- arcsin x is the **angle** in the interval $-\dfrac{\pi}{2}$ to $\dfrac{\pi}{2}$ for which the sine of this angle $= x$, so if arcsin $x = y$ then sin $y = x$.

- the domain of the function $y = \arcsin x$ is $\{x: -1 \leq x \leq 1\}$.

- the range of the function arcsin x is $\left\{y: -\dfrac{\pi}{2} \leq y \leq \dfrac{\pi}{2}\right\}$.

Essential notes

The relationship between the graph of a function and its inverse which is a reflection in the line $y = x$, was covered in Core 2.

Example
Without using a calculator, work out the value of the following giving your answers in degrees and radians i.e. in terms of π.

a) arcsin 0

b) $\arcsin\left(\dfrac{1}{2}\right)$

Continued on the next page

c) arcsin (-1)

d) arcsin $\left(-\dfrac{1}{\sqrt{2}}\right)$

e) arcsin $\left(\sin\left(-\dfrac{\pi}{5}\right)\right)$

Answer

a) For the inverse function to be defined remember that the domain is restricted to $\left\{x\colon\ -\dfrac{\pi}{2}\le x\le\dfrac{\pi}{2}\right\}.$

arcsin 0 means the angle whose sine is 0

Let this angle be x:

arcsin $0 = x$ means $\sin x = 0$

$\Rightarrow x = 0°$ or 0^c

b) arcsin $\left(\dfrac{1}{2}\right)$ means the angle whose sine is $\left(\dfrac{1}{2}\right)$.

Let this angle be x: arcsin $\left(\dfrac{1}{2}\right) = x$ means $\sin x = \left(\dfrac{1}{2}\right)$

but $\sin 30° = \left(\dfrac{1}{2}\right)$ so $x = 30°$ or $\dfrac{\pi}{6}$

c) arcsin (-1) means the angle whose sine is (-1).

Let this angle be x: arcsin $(-1) = x$ means $\sin x = (-1)$

but $\sin(-90°) = -1$ so $x = -90°$ or $-\dfrac{\pi}{2}$

d) arcsin $\left(-\dfrac{1}{\sqrt{2}}\right)$ means the angle whose sine is $\left(-\dfrac{1}{\sqrt{2}}\right)$.

Let this angle be x: arcsin $\left(-\dfrac{1}{\sqrt{2}}\right) = x$ means $\sin x = \left(-\dfrac{1}{\sqrt{2}}\right)$.

But $\sin(-45°) = \left(-\dfrac{1}{\sqrt{2}}\right)$ so $x = -45°$ or $-\dfrac{\pi}{4}$

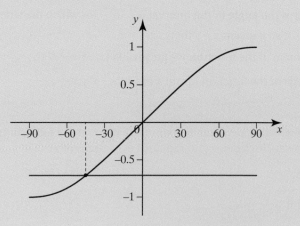

Method notes

Use the rule that arcsin x is the angle θ for which

$\sin\theta = x$. If the question says without using a calculator this means the angles will be common angles and you should learn their trigonometric ratios.

$\sin 30° = \dfrac{1}{2}$ or $\sin\dfrac{\pi}{6} = \dfrac{1}{2}$

$\sin 45° = \dfrac{1}{\sqrt{2}}$

A graph of $y = \sin x$ helps to explain the negative value in (d).

Fig. 2.14

Stop and think 1

Without using a calculator find the value of arcsin $\left(\sin\left(-\dfrac{\pi}{5} \right) \right)$. Give reasons for your answer.

Graph and properties of arccos x

To define the inverse function arccos x the domain of cos x is restricted to $\{x: 0 \le x \le \pi\}$. Cos x is then a one-one function in this restricted domain. Figure 2.15 shows the graphs of $y = \cos x$ (red) and $y = \arccos x$ (blue).

Essential notes

The relationship between the graph of a function and its inverse which is a reflection in the line $y = x$, was covered in Core 2.

Fig. 2.15
The graphs of $y = \cos x$ (red) and $y = \arccos x$ (blue).

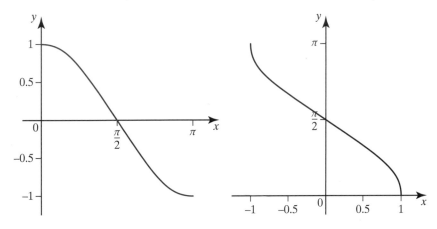

- The domain of arccos x is $\{x: -1 \le x \le 1\}$.

- Arccos x is the **angle** in the interval 0 to π for which the cos of this angle is x, so if arccos $x = \theta$ then $\cos\theta = x$.

- The range of arccos x is $\{y: 0 \le y \le \pi\}$.

Graph and properties of arctan x

To define the inverse function arctan x the domain of tan x is restricted to $\left\{x: -\dfrac{\pi}{2} \le x \le \dfrac{\pi}{2}\right\}$ so that tan x is then a one-one function in this restricted domain. Figure 2.16 shows the graphs of $y = \tan x$ (red) and $y = \arctan x$ (blue).

Fig. 2.16
The graphs of $y = \tan x$ (red) and $y = \arctan x$ (blue).

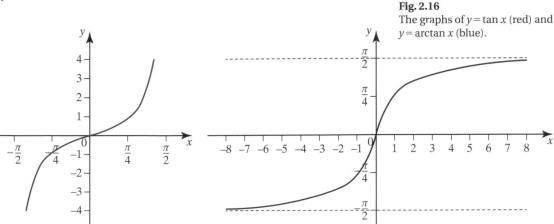

- The domain of arctan x is $\{x: -\infty \le x \le \infty\}$

- Arctan x is the **angle** in the interval $-\dfrac{\pi}{2}$ to $\dfrac{\pi}{2}$ for which the tangent of this angle is x, so if arctan $x = y$ then tan $y = x$.

- The range of arctan x is $\left\{x: -\dfrac{\pi}{2} \le x \le \dfrac{\pi}{2}\right\}$.

- The graph of $y =$ arctan x has horizontal asymptotes at
  $$y = -\frac{\pi}{2} \text{ and } y = \frac{\pi}{2}.$$

Compound angle formulae

In studying trigonometry you have become familiar with various relationships between trigonometric functions such as:

$$\sin^2 \theta + \cos^2 \theta \equiv 1$$

$$\tan\theta \equiv \frac{\sin\theta}{\cos\theta}$$

$$\sin(90° - \theta) \equiv \cos \theta$$

In each of these examples the identities are in terms of one variable θ. There are other useful relationships involving trigonometric functions of two variables (θ and ϕ) such as $\sin(\theta + \phi)$ and $\cos(\theta - \phi)$. These identities are often known as the **compound angle formulae.**

Example

Using a calculator complete the entries in the following tables:

θ	ϕ	$\sin(\theta)$	$\sin(\phi)$	$\sin(\theta) + \sin(\phi)$	$\sin(\theta + \phi)$
10°	57°				
15°	70°				
25°	45°				

θ	ϕ	$\cos(\theta)$	$\cos(\phi)$	$\cos(\theta) - \cos(\phi)$	$\cos(\theta - \phi)$
10°	57°				
15°	70°				
25°	45°				

Answer

θ	ϕ	$\sin(\theta)$	$\sin(\phi)$	$\sin(\theta) + \sin(\phi)$	$\sin(\theta + \phi)$
10°	57°	0.1736	0.8387	1.0123	0.9205
15°	70°	0.2588	0.9397	1.1985	0.9962
25°	45°	0.4226	0.7071	1.1297	0.9397

θ	ϕ	$\cos(\theta)$	$\cos(\phi)$	$\cos(\theta) - \cos(\phi)$	$\cos(\theta - \phi)$
10°	57°	0.9848	0.5446	0.4402	0.6820
15°	70°	0.9659	0.3420	0.6239	0.5736
25°	45°	0.9063	0.7071	0.1992	0.9397

It is important to notice from the entries in these tables that, in general:

$\sin(\theta + \phi) \neq \sin\theta + \sin\phi$ and

$\cos(\theta - \phi) \neq \cos\theta - \cos\phi$

Proving the formula for $\cos(\theta - \phi)$

The diagram in Figure 2.17 shows three points O, A and B where O is the origin and A and B are such that $OA = 1$ and $OB = 1$

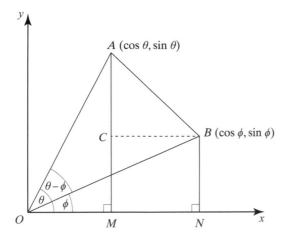

- OA makes an angle θ with the x-axis and OB makes an angle ϕ with the x-axis.
- AM and BN are perpendicular lines from points A and B to the x-axis respectively.
- OAM forms a right angled triangle, so the lengths of OM and AM are $\cos\theta$ and $\sin\theta$ respectively. Thus the coordinates of A are $(\cos\theta, \sin\theta)$.
- Similarly, OBN forms a right angled triangle, so the lengths of ON and BN are $\cos\phi$ and $\sin\phi$ respectively. Thus the coordinates of B are $(\cos\phi, \sin\phi)$.

Apply the cosine rule to the triangle OAB:

$$AB^2 \equiv OA^2 + OB^2 + 2 \times OA \times OB \times \cos(\theta - \phi)$$

$$\equiv 1 + 1 - 2\cos(\theta - \phi)$$

$$\equiv 2 - 2\cos(\theta - \phi) \qquad (1)$$

Method notes

To complete $\sin(\theta) + \sin(\phi)$ use your calculator to work out $\sin(\theta)$ and $\sin(\phi)$ for the given values of θ and ϕ then add together the individual sine values.

To complete $\sin(\theta + \phi)$ use your calculator to add together θ and ϕ for the two given values of θ and ϕ then use the sine button to work out the sine of the angle $(\theta + \phi)$.

Fig. 2.17
Diagram to derive the formula for $\cos(\theta - \phi)$.

Method notes

In triangle OAM:

$\angle AMO = 90°$

$\Rightarrow \dfrac{OM}{OA} = \cos\theta$

$OA = 1 \Rightarrow OM = \cos\theta$

In triangle OAM:

$\dfrac{AM}{OA} = \sin\theta$

$OA = 1 \Rightarrow AM = \sin\theta$

Similarly triangle OBN is right angled at N and $OB = 1$

$\Rightarrow ON = \cos\phi$ and

$BN = \sin\phi$

The cosine rule was introduced in Core 2 as:

$a^2 \equiv b^2 + c^2 + 2ab\cos A$

using the conventional labelling for sides, so that a is the side opposite $\angle A$.

Apply Pythagoras' Theorem to triangle ABC:

$$AB^2 \equiv BC^2 + AC^2 \equiv (\cos\phi - \cos\theta)^2 + (\sin\theta - \sin\phi)^2$$

$$\equiv (\cos^2\phi - 2\cos\theta\cos\phi + \cos^2\theta) + (\sin^2\theta - 2\sin\theta\sin\phi + \sin^2\phi)$$

$$\equiv 2 - 2\cos\theta\cos\phi - 2\sin\theta\sin\phi \qquad (2)$$

Equating expressions (1) and (2)

$$\Rightarrow 2 - 2\cos(\theta - \phi) \equiv 2 - 2\cos\theta\cos\phi - 2\sin\theta\sin\phi$$

$$\Rightarrow \qquad \cos(\theta - \phi) \equiv \cos\theta\cos\phi + \sin\theta\sin\phi$$

The formula can be given in terms of any two angles so in terms of angles A and B it is:

$$\cos(A - B) = \cos A \cos B + \sin A \sin B$$

Deriving the formula for $\cos(\theta + \phi)$

Starting with the formula just proved:

$$\cos(\theta - \phi) \equiv \cos\theta\cos\phi + \sin\theta\sin\phi$$

and replacing ϕ by $-\phi$ gives

$$\cos(\theta - (-\phi)) \equiv \cos\theta\cos(-\phi) + \sin\theta\sin(-\phi)$$

$$\cos(\theta + \phi) \equiv \cos\theta\cos\phi - \sin\theta\sin\phi$$

or in terms of $\angle A$ and $\angle B$:

$$\cos(A + B) \equiv \cos A \cos B - \sin A \sin B$$

Deriving the formulae for $\sin(\theta + \phi)$ and $\sin(\theta - \phi)$

Compound formulas for $\sin(\theta + \phi)$ and $\sin(\theta - \phi)$ can be derived from the expressions for $\cos(\theta + \phi)$ and $\cos(\theta - \phi)$.

Deriving the formula for $\sin(\theta + \phi)$

Step 1: $\sin(\theta + \phi) \equiv \cos\left(\dfrac{\pi}{2} - (\theta + \phi)\right)$

$$\equiv \cos\left(\left(\dfrac{\pi}{2} - \theta\right) - \phi\right)$$

Step 2: Use the compound angle formula

$$\cos(A - B) = \cos A \cos B + \sin A \sin B$$

in step 1 with $\angle A = \left(\dfrac{\pi}{2} - \theta\right)$ and $\angle B = \phi$

$$\Rightarrow \cos\left(\left(\dfrac{\pi}{2} - \theta\right) - \phi\right) \equiv \cos\left(\dfrac{\pi}{2} - \theta\right)\cos\phi + \sin\left(\dfrac{\pi}{2} - \theta\right)\sin\phi$$

$$\equiv \sin\theta\cos\phi + \cos\theta\sin\phi$$

Step 3: Use the result of step 2 in step 1

$$\Rightarrow \sin(\theta + \phi) \equiv \sin\theta\cos\phi + \cos\theta\sin\phi$$

Deriving the formula for sin(θ − φ)

Step 1: $\sin(\theta - \phi) \equiv \cos\left(\dfrac{\pi}{2} - (\theta - \phi)\right) \equiv \cos\left(\left(\dfrac{\pi}{2} - \theta\right) + \phi\right)$

Step 2: Use the compound angle formula

$$\cos(A + B) \equiv \cos A \cos B - \sin A \sin B$$

in step 1 with $\angle A = \left(\dfrac{\pi}{2} - \theta\right)$ and $\angle B = \phi$

$$\Rightarrow \cos\left(\left(\dfrac{\pi}{2} - \theta\right) + \phi\right) \equiv \cos\left(\dfrac{\pi}{2} - \theta\right)\cos\phi - \sin\left(\dfrac{\pi}{2} - \theta\right)\sin\phi$$

$$\equiv \sin\theta\cos\phi - \cos\theta\sin\phi$$

Step 3: Use the result of step 2 in step 1

$$\Rightarrow \sin(\theta - \phi) \equiv \sin\theta\cos\phi - \cos\theta\sin\phi$$

Proving further trigonometric identities

Example
Prove that

a) $\tan(A + B) \equiv \dfrac{\tan A + \tan B}{1 - \tan A \tan B}$

b) $\tan(A - B) \equiv \dfrac{\tan A - \tan B}{1 + \tan A \tan B}$

Answer
a) **Step 1:** Start with the LHS of the identity and rewrite using the trigonometric identity:

$$\tan(A + B) \equiv \dfrac{\sin(A + B)}{\cos(A + B)}$$

Step 2: Rewrite $\dfrac{\sin(A + B)}{\cos(A + B)}$ using the compound angle formulae

$$\Rightarrow \dfrac{\sin(A + B)}{\cos(A + B)} \equiv \dfrac{\sin A \cos B + \cos A \sin B}{\cos A \cos B - \sin A \sin B}$$

Step 3: Divide the numerator of the fraction in the RHS of the identity from step 2 by $\cos A \cos B$:

$$\left(\dfrac{\sin A \cos B}{\cos A \cos B} + \dfrac{\cos A \sin B}{\cos A \cos B}\right) = \dfrac{\sin A}{\cos A} + \dfrac{\sin B}{\cos B}$$

Therefore the numerator of the fraction from step 2

$$= \tan A + \tan B$$

Continued on the next page

Step 4: Divide the denominator of the fraction in the RHS of the identity from step 2 by

$$\cos A \cos B \Rightarrow \left(\frac{\cos A \cos B}{\cos A \cos B} - \frac{\sin A \sin B}{\cos A \cos B} \right) = 1 - \tan A \tan B$$

Step 5: Use the results of step 3 and step 4 in step 1

$$\Rightarrow \tan(A + B) \equiv \frac{\tan A + \tan B}{1 - \tan A \tan B}$$

Therefore the identity is proved.

b) **Step 1:** Start with the LHS of the identity and rewrite using the

trigonometric identity $\tan(A - B) \equiv \dfrac{\sin(A - B)}{\cos(A - B)}$

Step 2: Rewrite $\dfrac{\sin(A - B)}{\cos(A - B)}$ using the compound angle formulae for $\sin(A - B)$ and $\cos(A - B)$ so:

$$\frac{\sin(A - B)}{\cos(A - B)} \equiv \frac{\sin A \cos B - \cos A \sin B}{\cos A \cos B + \sin A \sin B}$$

Step 3: Divide each term of the fraction in the answer of step 2 by

$\cos A \cos B$ so the numerator is $\dfrac{\sin A}{\cos A} - \dfrac{\sin B}{\cos B}$

$$= \tan A - \tan B$$

and the denominator is $1 + \dfrac{\sin A \sin B}{\cos A \cos B}$

$$= 1 + \tan A \tan B$$

Step 4: Simplify the complete fraction in step 2 using the answers

from step 3 $\equiv \dfrac{\tan A - \tan B}{1 + \tan A \tan B} = \text{RHS}$

Therefore the identity is proved.

Method notes

Exact means give answers in terms of surds. It also means that common angles will be involved in the question.

15° is not a common angle but to rewrite it in terms of common angles $\Rightarrow$ 15° = (60° − 45°) then

sin(60° − 45°) can be rewritten using the compound angle formula.

Example

Without using a calculator find the exact value of

a) sin 15° and b) tan 105°.

Answer

a) $\sin 15° = \sin(60° - 45°)$

$$= \sin 60° \cos 45° - \cos 60° \sin 45°$$

$$= \frac{\sqrt{3}}{2} \times \frac{\sqrt{2}}{2} - \frac{1}{2} \frac{\sqrt{2}}{2} = \frac{\sqrt{6} - \sqrt{2}}{4}$$

b) $\tan 105° = \tan(60° + 45°)$

$$= \frac{\tan 60 + \tan 45}{1 - \tan 60 \tan 45}$$

$$= \frac{\sqrt{3} + 1}{1 - \sqrt{3} \times 1} = \frac{\sqrt{3} + 1}{1 - \sqrt{3}} = \frac{-(1 + \sqrt{3})^2}{2}$$

Method notes

b) $105° = (60° + 45°)$.

Example

Without using a calculator, find the exact value of

a) $\cos(A + B)$

b) $\sin(A - B)$ given that A and B are both acute angles and $\tan A = \dfrac{4}{3}$

and $\sin B = \dfrac{5}{13}$

Fig. 2.18

Answer

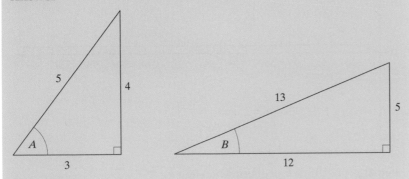

From these triangles since A and B are acute, sine and cosine will be positive:

$$\sin A = \frac{4}{5}, \quad \cos A = \frac{3}{5}, \quad \cos B = \frac{12}{13}$$

a) $\cos(A + B) = \cos A \cos B - \sin A \sin B$

$$= \frac{3}{5} \times \frac{12}{13} - \frac{4}{5} \times \frac{5}{13} = \frac{16}{65}$$

b) $\sin(A - B) = \sin A \cos B - \cos A \sin B$

$$= \frac{4}{5} \times \frac{12}{13} - \frac{3}{5} \times \frac{5}{13} = \frac{33}{65}$$

Method notes

To answer the question we need to know the values of cos A, sin A and cos B as these appear in the compound angle formulae for (a) and (b).

Draw a right angled triangle with an angle A in it. Given that $\tan A = \dfrac{4}{3}$ the side opposite A is 4 and the side adjacent to A is 3

Use Pythagoras Theorem to find the hypotenuse length:

$4^2 + 3^2 = 16 + 9$

$\qquad = 25$

$\qquad = (\text{hypotenuse})^2$

$\qquad \Rightarrow \text{hypotenuse} = 5$

Draw another right angled triangle with an angle B in it. Given that $\sin B = \dfrac{5}{13}$ use Pythagoras to find the length of the third side in this triangle $= 12$

☞ Continued on the next page

Method notes

$$\frac{\sin x}{\cos x} = \tan x$$

Fig. 2.19
Graphs of $y = \tan x$ and $y = 2 - \sqrt{3}$

Example
Solve the equation $\sin(x + 60°) = \cos x$ for $0 \le x \le 360°$.

Answer
Step 1: Expand $\sin(x + 60°)$ using the compound angle formula
$$\Rightarrow \sin x \cos 60° + \cos x \sin 60° = \cos x$$

Step 2: Divide by $\cos x \Rightarrow \tan x \cos 60° + \sin 60° = 1$

Step 3: Simplify $\Rightarrow \tan x = \dfrac{1 - \sin 60°}{\cos 60°} = \dfrac{1 - \dfrac{\sqrt{3}}{2}}{\dfrac{1}{2}} = 2 - \sqrt{3}$

Step 4: Draw the graph of $\tan x$ and the line $2 - \sqrt{3}$

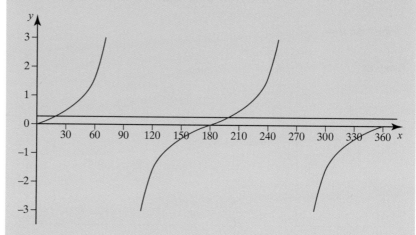

Step 5: $x = 15°$ (from calculator where $\tan x = 2 - \sqrt{3}$) and
$$x = 180° + 15° = 195° \text{ (from the graph)}$$
These are the x-values where the curve and the line cross.

Double angle formulae
Special results known as the double angle formulae arise from the compound angle formulae when A and B are equal.
If $B = A$, $\sin(A + B) \equiv \sin(A)\cos(B) + \cos(A)\sin(B)$ so:

$$\sin(A + A) \equiv \sin(A)\cos(A) + \cos(A)\sin(A) \equiv 2\sin(A)\cos(A)$$

$$\Rightarrow \mathbf{\sin 2A \equiv 2\sin A \cos A}$$

which is the double angle formula for sine.

If $B = A$, $\cos(A + B) = \cos(A)\cos(B) - \sin(A)\sin(B)$ so:

$$\cos(A + A) \equiv \cos(A)\cos(A) - \sin(A)\sin(A) \equiv \cos^2(A) - \sin^2(A)$$

$$\Rightarrow \mathbf{\cos 2A \equiv \cos^2 A - \sin^2 A}$$

which is the double angle formula for cosine.

If we then use the identity $\cos^2 A + \sin^2 A \equiv 1$

$\Rightarrow \sin^2 A \equiv 1 - \cos^2 A$

Use this to replace $\sin^2 A$ in $\cos 2A \equiv \cos^2 A - \sin^2 A$

$\Rightarrow \qquad \cos 2A \equiv \cos^2 A - (1 - \cos^2 A) \equiv 2\cos^2 A - 1$

$\Rightarrow \qquad \mathbf{\cos 2A \equiv 2\cos^2 A - 1}$

which is another double angle formula for cosine.

But we can also rewrite $\cos^2 A + \sin^2 A \equiv 1$ as $\cos^2 A \equiv 1 - \sin^2 A$

Use this to replace $\cos^2 A$ in $\cos 2A \equiv \cos^2 A - \sin^2 A$

$\Rightarrow \qquad \cos 2A \equiv (1 - \sin^2 A) - \sin^2 A \equiv 1 - 2\sin^2 A$

$\Rightarrow \qquad \mathbf{\cos 2A \equiv 1 - 2\sin^2 A}$

which is another double angle formula for cosine.

If $B = A$, in $\tan(A + B) \equiv \dfrac{\tan A + \tan B}{1 - \tan A \tan B}$ and

$\tan(A + A) \equiv \dfrac{\tan A + \tan A}{1 - \tan A \tan A} = \dfrac{2\tan A}{1 - \tan^2 A}$

$\Rightarrow \qquad \mathbf{\tan 2A \equiv \dfrac{2\tan A}{1 - \tan^2 A}}$

which is a double angle formula for tangent.

Exam tips
You must learn all 5 of the double angle formulae as they are not in the formula booklet. You are likely to be tested in the examination on the application of these formulae for problem solving.

Stop and think 2

Using the double angle formulae and without using a calculator find the values of :

a) $\sin 120°$

b) $\cos \pi^c$

c) $\tan 90°$

Example

Given that $\cos \theta = 0.56$ and θ is acute, find the value of each of the following correct to 2 d.p.:

a) $\sin \theta$

b) $\sin 2\theta$

c) $\cos 2\theta$

d) $\tan 2\theta$

☞ Continued on the next page

Method notes

a) Use the double angle formula for sine and substitute the known values for sin θ and cos θ.

b) As the value of cos θ was given in the question use the double angle formula for cosine which involves only cos θ not sin θ.

Answer

a) **Step 1:** Given the value of $\cos\theta$, to find $\sin\theta$ use the identity

$$\sin^2\theta + \cos^2\theta \equiv 1$$

$$\Rightarrow \quad \sin^2\theta \equiv 1 - \cos^2\theta$$

$$\Rightarrow 1 - (0.56)^2 = 0.6864$$

Step 2: Take the square root which gives $\sin\theta = +0.83$ or -0.83

Step 3: Given that θ is acute we can reject $\sin\theta = -0.83$ and the solution is $\theta = 0.83$

b) $\sin 2\theta \equiv 2\sin\theta\cos\theta = 2 \times 0.56 \times 0.83 = 0.93$

c) $\cos 2\theta \equiv 2\cos^2\theta - 1 = 2 \times 0.56^2 - 1 = -0.37$

d) **Step 1:** The double angle formula for tan involves $\tan\theta$ so we need to know the value of $\tan\theta$. Use values from (a):

$$\tan\theta = \frac{\sin\theta}{\cos\theta} = \frac{0.83}{0.56} = 1.482$$

Step 2: Using the double angle formula gives:

$$\tan 2\theta = \frac{2\tan\theta}{1 - \tan^2\theta} = \frac{2 \times 1.482}{1 - 1.482^2} = -2.48$$

Example

Show that $\dfrac{\cos 2A}{\cos A + \sin A} \equiv \cos A - \sin A$

Answer

Step 1: Start with the LHS of the identity and rewrite $\cos 2A$ using the double angle formula: $\dfrac{\cos 2A}{\cos A + \sin A} \equiv \dfrac{\cos^2 A - \sin^2 A}{\cos A + \sin A}$

Step 2: Factorise $\cos^2 A - \sin^2 A$ using the difference of two squares:

$$\frac{(\cos A - \sin A)(\cos A + \sin A)}{\cos A + \sin A} \equiv \cos A - \sin A = \text{RHS of the}$$

identity.

Therefore the identity is proved.

Example

a) Show that $\sin 3\theta \equiv 3\sin\theta - 4\sin^3\theta$

b) Hence solve $8\sin^3\theta - 6\sin\theta + 1 = 0$ for $0 < \theta < \dfrac{\pi}{3}$, giving your answers in terms of π.

Answer

a) **Step 1:** To find an expression for $\sin 3\theta$, put $\phi = 2\theta$ into

$$\sin(\theta + \phi) \equiv \sin\theta\cos\phi + \cos\theta\sin\phi$$
$$\Rightarrow \sin(\theta + 2\theta) \equiv \sin\theta\cos 2\theta + \cos\theta\sin 2\theta$$

Step 2: Use the double angle formulae for $\sin 2\theta$ and $\cos 2\theta$ in step 1:

$$\sin(3\theta) \equiv \sin\theta\,(1 - 2\sin^2\theta) + \cos\theta\,(2\sin\theta\cos\theta)$$
$$\equiv \sin\theta\,(1 - 2\sin^2\theta + 2\cos^2\theta)$$

Step 3: Simplify using the Pythagorean Identity:

$$\equiv \sin\theta\,(1 - 2\sin^2\theta + 2\,(1 - \sin^2\theta\,))$$
$$\equiv \sin\theta\,(3 - 4\sin^2\theta)$$
$$\equiv 3\sin\theta - 4\sin^3\theta = \text{RHS of the identity.}$$

Therefore the identity is shown.

b) 'Hence' implies that the result in a) will be useful:

Step 1: Factorise $8\sin^3\theta - 6\sin\theta + 1 = 0$

$$\Rightarrow \quad 2(4\sin^3\theta - 3\sin\theta) + 1 = 0$$

Step 2: From a) $\sin 3\theta \equiv 3\sin\theta - 4\sin^3\theta$

From step 1: $2(4\sin^3\theta - 3\sin\theta) + 1 = -2\sin 3\theta + 1 = 0$

Therefore $\sin 3\theta = \dfrac{1}{2}$

Step 3: Draw the graphs $y = \sin x$ (where $x = 3\theta$) and $y = \dfrac{1}{2}$

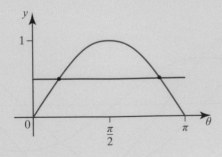

Step 4: From the calculator $3\theta = \dfrac{\pi}{6} \Rightarrow \theta = \dfrac{\pi}{18}$ or

$$3\theta = \pi - \dfrac{\pi}{6} = \dfrac{5\pi}{6} \Rightarrow \theta = \dfrac{5\pi}{18}$$

Fig. 2.20

Graphs of $y = \sin x$ and $y = \dfrac{1}{2}$

Wave functions

The graphs of the trigonometric functions sine and cosine have 'wave like' features and can be used to describe physical situations with periodic features.

Figure 2.21 shows the graphs of $y = 3 \sin x$ and $y = 4 \cos x$ for the interval $0 \le x \le 360°$.

Fig. 2.21
Graphs of $y = 3 \sin x$ (red) and $y = 4 \cos x$ (blue).

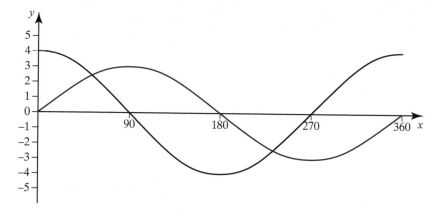

- Each graph is periodic with **period** 360°.

- The maximum value of $y = 3 \sin x$ is 3 and the minimum value is −3

- The maximum value of $y = 4 \cos x$ is 4 and the minimum value is −4

- The values 3 and 4 are called the **amplitude** of the waves.

- By comparing the peaks of each graph, you can see that the sine graph $y = 3 \sin x$ is translated parallel to the x-axis by 90° from the cosine graph. We say that the sine graph is **out of phase** with the cosine graph by 90°.

Combining wave functions

Figure 2.22 shows the graph of the addition of the two waves $y = 4 \cos x$ and $y = 3 \sin x$.

Fig. 2.22
Graphs of $y = 3 \sin x$ (red), $y = 4 \cos x$ (blue) and $y = 4 \cos x + 3 \sin x$ (purple).

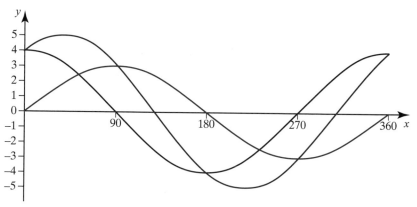

The graph of $y = 4 \cos x + 3 \sin x$ is a wave with period 360° and amplitude 5. This 'combination' graph has a 'wave like' feature so we can try to fit it to a cosine function or a sine function.

Suppose that we try to fit $y = 4 \cos x + 3 \sin x$ to a cosine function of the form $y = 5 \cos(x - \alpha)$ where α is an acute angle. The factor 5 is needed because we know the amplitude of the combination is 5

Step 1: Let $5 \cos(x - \alpha) \equiv 4 \cos x + 3 \sin x$

Step 2: Use the compound angle formula for $\cos(x - \alpha)$:

$$5 \cos(x - \alpha) \equiv 5 \cos x \cos \alpha + 5 \sin x \sin \alpha$$

Step 3: Use this to rewrite the equation in step 1:

$$5 \cos x \cos \alpha + 5 \sin x \sin \alpha \equiv 4 \cos x + 3 \sin x$$

Step 4: Equate the coefficients of $\cos x$ in the identity of step 3 so $(5 \cos \alpha = 4)$

and equate the coefficients of $\sin x$ in the identity of step 3 so $(5 \sin \alpha = 3)$.

Step 5: Solve both of the equations in step 4. They must give the same answer otherwise the original statement in step 1 cannot be true.

$$5 \cos \alpha = 4 \Rightarrow \alpha = \arccos 0.8 \Rightarrow \alpha = 53°$$

$$5 \sin \alpha = 3 \Rightarrow \alpha = \arcsin 0.6 \Rightarrow \alpha = 53°$$

Step 6: Rewrite the equation in step 1:

$$4 \cos x + 3 \sin x \equiv 5 \cos(x - 53°)$$

This shows that the 'combination' function $y = 4 \cos x + 3 \sin x$ of the two wave functions is the same as the function

$$y = 5 \cos(x - 53°).$$

Essential notes

In this example the amplitude was easy to read from the graph.

But $5^2 = 4^2 + 3^2$ so the square of the amplitude of $y = 4 \cos x + 3 \sin x$ is the sum of the squares of the separate amplitudes of the functions

$y = 4 \cos x$ and $y = 3 \sin x$.

Example
Express $5 \cos x + 12 \sin x$ in the form $R \cos(x - \alpha)$ where R is a positive constant and $\angle \alpha$ is acute.

Answer
Step 1: Let $R \cos(x - \alpha) \equiv 5 \cos x + 12 \sin x$.

Step 2: Expand $R \cos(x - \alpha)$ using the addition formula:

$$R \cos(x - \alpha) \equiv R \cos x \cos \alpha + R \sin x \sin \alpha$$

Step 3: Use the answer from step 2 to rewrite the identity in step 1:

$$R \cos x \cos \alpha + R \sin x \sin \alpha \equiv 5 \cos x + 12 \sin x$$

☞ Continued on the next page

Step 4: Compare the coefficients of $\cos x$ and $\sin x$ in the identity in step 3:

$R\cos\alpha = 5$ and $R\sin\alpha = 12$

Step 5: Square both equations from step 4 and then add them together:

$$R^2\cos^2\alpha + R^2\sin2\,\alpha = 5^2 + 12^2$$
$$\Rightarrow R^2(\cos^2\alpha + \sin^2\alpha) = 25 + 144 = 169$$
$$\Rightarrow \qquad\qquad R^2 = 169$$
$$\Rightarrow \qquad\qquad R = 13 \text{ or } -13$$

but we can reject the value $R = -13$ as R was given as a positive constant.

Step 6: Divide out the two equations from step 4:

$$\frac{R\sin\alpha}{R\cos\alpha} = \tan\alpha = \frac{12}{5}$$
$$\Rightarrow \qquad\qquad \alpha = \arctan\frac{12}{5} \Rightarrow \alpha = 67°$$

α was given as acute so we do not need to consider any of the other values where $\tan\alpha = \dfrac{12}{5}$

Step 7: Rewrite the identity in step 1 using the values we have found for R and α:

$$5\cos x + 12\sin x \equiv 13\cos(x - 67°)$$

In the last two examples we fitted the 'addition' graph to the cosine function. We could have chosen to fit it to the sine function. The following example will explore this further.

Solution of equations using combined wave functions

Example

a) Express $4\sin x - 3\cos x$ in the form $R\sin(x - \alpha)$ where R is a positive constant and $\angle\alpha$ is acute.

b) Hence solve the equation $4\sin x - 3\cos x = 1$ for $0 < x < 360°$.

Answer

a) **Step 1:** Let $R\sin(x - \alpha) \equiv 4\sin x - 3\cos x$

Step 2: Expand $R\sin(x - \alpha)$ using the addition formula:

$$R\sin(x - \alpha) \equiv R\sin x\cos\alpha - R\cos x\sin\alpha$$

Step 3: Use the answer from step 2 to rewrite the identity in step 1:

$$R\sin x\cos\alpha - R\cos x\sin\alpha \equiv 4\sin x - 3\cos x$$

Step 4: Compare the coefficients of $\sin x$ and $\cos x$ in the identity in step 3:

$$R\cos\alpha = 4 \text{ and } R\sin\alpha = 3$$

Step 5: Square both equations in step 4 and add them together:

$$R^2 \cos^2 \alpha + R^2 \sin^2 \alpha = 4^2 + 3^2 = 25$$

$$\Rightarrow R^2 (\cos^2 \alpha + \sin^2 \alpha) = 16 + 9 = 25$$

$$\Rightarrow \qquad\qquad R^2 = 25$$

$$\Rightarrow \qquad\qquad R = 5 \text{ or } R = -5$$

but we can reject the value -5 as R was given as a positive constant.

Step 6: Divide the equations from Step 4:

$$\frac{R \sin \alpha}{R \cos \alpha} = \tan \alpha = \frac{3}{4}$$

$$\Rightarrow \qquad \alpha = \arctan \frac{3}{4}$$

$$\Rightarrow \qquad \alpha = 36.9°$$

α was given as acute so we do not need to consider any of the other values where $\tan \alpha = \dfrac{3}{4}$

Step 7: Rewrite the identity in step 1 using the values we have found for R and α:

$$4 \sin x - 3 \cos x \equiv 5 \sin(x - 36.9°)$$

b) 'Hence' means that the answer from a) will be helpful in solving the equation:

$$4 \sin x - 3 \cos x = 1 \text{ for } 0 < x < 360°.$$

Step 1: Use the answer from a) to rewrite the equation we are asked to solve:

$$5 \sin(x - 36.9°) = 1$$

$$\Rightarrow \sin(x - 36.9°) = \frac{1}{5}$$

Step 2: Use a calculator to find all the values of x such that:

$$\sin(x - 36.9°) = \frac{1}{5}$$

$$\Rightarrow \quad x - 36.9° = \arcsin \frac{1}{5}$$

$$\Rightarrow \quad x - 36.9° = 11.5° \text{ or } x - 36.9° = 180° - 11.5° = 168.5°$$

$$\Rightarrow \qquad\qquad x = 36.9° + 11.5° = 48.4° \text{ or } x = 205.4°$$

are the solutions of the equation for $0 < x < 360°$.

Method notes

In this question, x is given in terms of π which means you must use your calculator in radian mode.

Example

Solve the equation $2 \sin x - 3 \cos x = 2$ for $0 < x < 2\pi$.

Answer

This equation involves two wave functions so we know that the method of solution will be to 'fit' the combination function to a sine or a cosine function.

The two wave functions are subtracted therefore we usually choose to rewrite this as a sine function.

Step 1: Let $R \sin(x - \alpha) \equiv 2 \sin x - 3 \cos x$ where R is positive and α is an acute angle.

Step 2: Expand $R \sin(x - \alpha)$ in step 1 using the compound angle formula:

$R \sin x \cos \alpha - R \cos x \sin \alpha \equiv 2 \sin x - 3 \cos x$

Step 3: Compare the coefficients of $\sin x$ and $\cos x$ of the identity in step 2:

$R \cos \alpha = 2$ and $R \sin \alpha = 3$

Step 4: Square both equations in step 3 and add them together:

$R^2 = 2^2 + 3^2 = 13$ because $R^2 \cos^2 \alpha + R^2 \sin^2 \alpha = R^2$

$\Rightarrow R = \sqrt{13}$ and we can reject the negative value

Step 5: Divide out both equations in step 3:

$$\frac{R \sin \alpha}{R \cos \alpha} = \tan \alpha = \frac{3}{2} \Rightarrow \alpha = \arctan \frac{3}{2} = 0.983^c$$

Step 6: Replace R and α in step 1:

$2 \sin x - 3 \cos x = \sqrt{13} \sin(x - 0.983)$

The original equation can now be written as:

$\sqrt{13} \sin(x - 0.983) = 2$

$$\Rightarrow \quad \sin(x - 0.983) = \frac{2}{\sqrt{13}} = 0.5547$$

$\Rightarrow x - 0.983 = \arcsin 0.5547 = 0.588^c$ (primary solution)

or $x - 0.983 = \pi - 0.588 = 2.554^c$ (secondary solution)

$\Rightarrow x = 1.571^c$ or $x = 3.537^c$

are the solutions of the equation $2 \sin x - 3 \cos x = 2$ for $0 < x < 2\pi$.

Proving the factor formulae for adding sines and cosines

Example

Prove that $\sin P + \sin Q \equiv 2 \sin\left(\dfrac{P + Q}{2}\right)\cos\left(\dfrac{P - Q}{2}\right)$

Answer

We know from our knowledge of the compound angle formulae that:

$\sin(A + B) \equiv \sin A \cos B + \cos A \sin B$

$\sin(A - B) \equiv \sin A \cos B - \cos A \sin B$

Step 1: Add together these two formulae:

$\quad \sin(A + B) + \sin(A - B) \equiv 2 \sin A \cos B$

Step 2: Let $(A + B) = P$ and $(A - B) = Q$ in the identity of step 1:

$\quad \sin P + \sin Q \equiv 2 \sin A \cos B$

Step 3: Find A and B in terms of P and Q by adding the equations in step 2:

$\quad (A + B) + (A - B) = P + Q$

$\Rightarrow \qquad\qquad 2A = P + Q$

$\Rightarrow \qquad\qquad A = \dfrac{P + Q}{2}$

and by subtracting the equations in step 2

$\Rightarrow (A + B) - (A - B) = P - Q$

$\Rightarrow \qquad\qquad 2B = P - Q$

$\Rightarrow \qquad\qquad B = \dfrac{P - Q}{2}$

Step 4: Replace A and B in terms of P and Q in the identity of step 2

$\Rightarrow \sin P + \sin Q \equiv 2 \sin\left(\dfrac{P + Q}{2}\right)\cos\left(\dfrac{P - Q}{2}\right)$

This is one of four expressions involving the addition and subtraction of trigonometric functions of different angles. These are called the **trigonometric factor formulae**.

The four trigonometric factor formulae are:

- $\sin P + \sin Q \equiv 2 \sin\left(\dfrac{P+Q}{2}\right)\cos\left(\dfrac{P-Q}{2}\right)$

- $\sin P - \sin Q \equiv 2 \cos\left(\dfrac{P+Q}{2}\right)\sin\left(\dfrac{P-Q}{2}\right)$

- $\cos P + \cos Q \equiv 2 \cos\left(\dfrac{P+Q}{2}\right)\cos\left(\dfrac{P-Q}{2}\right)$

- $\cos P - \cos Q \equiv -2 \sin\left(\dfrac{P+Q}{2}\right)\sin\left(\dfrac{P-Q}{2}\right)$

Exam tips

You do not need to learn the factor formulae as they are in the formula booklet.

The following examples illustrate how the factor formulae can be used to prove identities and to solve equations.

Example

Prove that $\dfrac{\sin 75^\circ + \sin 15^\circ}{\cos 15^\circ - \cos 75^\circ} \equiv \cot 30^\circ$

Answer

Step 1: Start with the LHS of the identity and rewrite the fraction using the factor formulae:

$$\frac{\sin 75^\circ + \sin 15^\circ}{\cos 15^\circ - \cos 75^\circ} = \frac{2 \sin \dfrac{(75^\circ + 15^\circ)}{2} \cos \dfrac{(75^\circ - 15^\circ)}{2}}{-2 \sin \dfrac{(15^\circ + 75^\circ)}{2} \sin \dfrac{(15^\circ - 75^\circ)}{2}}$$

Step 2: Simplify the answer in step 1

$$= \frac{2 \sin 45^\circ \cos 30^\circ}{-2 \sin 45^\circ \sin(-30^\circ)}$$

Step 3: Simplify the fraction further using the fact that $\sin(30) = -\sin(30°)$

$$= \frac{2 \sin 45^\circ \cos 30^\circ}{-2 \sin 45^\circ \sin(-30^\circ)}$$

$$= \frac{\cos 30^\circ}{\sin 30^\circ}$$

$$= \cot 30^\circ = \text{RHS of the identity.}$$

Therefore the identity is proved.

Solving equations using the factor formulae

Example
Solve the equation $\sin 3x - \sin x = \cos 2x$, for $0 \leq x \leq 2\pi$.

Answer
Step 1: The angles in the LHS of the equation are different ($3x$ and x) therefore simplify using the factor formula

$$\sin P - \sin Q \equiv 2 \cos\left(\frac{P+Q}{2}\right)\sin\left(\frac{P-Q}{2}\right) \text{with } P = 3x \text{ and } Q = x:$$

$$\Rightarrow \sin 3x - \sin x = 2 \cos 2x \sin x$$

Step 2: Rewrite the original equation using the result from step 1:

$$2 \cos 2x \sin x = \cos 2x$$

Step 3: Factorise the equation in step 2 and then solve

$$\Rightarrow \cos 2x (2 \sin x - 1) = 0$$

$$\Rightarrow \cos 2x = 0 \textbf{ or } 2 \sin x - 1 = 0 \text{ for } x \text{ in the range } 0 \leq x \leq 2\pi$$

$$\Rightarrow \cos 2x = 0$$

$$\Rightarrow \quad 2x = \frac{\pi}{2}, \frac{3\pi}{2}, \frac{5\pi}{2}, \frac{7\pi}{2}...$$

$$\Rightarrow \quad x = \frac{\pi}{4}, \frac{3\pi}{4}, \frac{5\pi}{4}, \frac{7\pi}{4}...$$

$$\Rightarrow \quad x = \frac{\pi}{4} \text{ or } x = \frac{3\pi}{4}$$

or

$$\Rightarrow 2 \sin x - 1 = 0$$

$$\Rightarrow \quad \sin x = \frac{1}{2}$$

$$\Rightarrow \quad x = \frac{\pi}{6} \text{ or } x = \frac{5\pi}{6}$$

Therefore the six solutions are: $x = \dfrac{\pi}{4}, \dfrac{3\pi}{4}, \dfrac{5\pi}{4}, \dfrac{7\pi}{4}, \dfrac{\pi}{6}, \dfrac{5\pi}{6}$

Stop and think answers

1 To evaluate $\arcsin\left(\sin\left(\dfrac{-\pi}{5}\right)\right)$

Let $\arcsin X = \alpha$ (1)

and comparing this with $\arcsin\left(\sin\left(\dfrac{-\pi}{5}\right)\right)$ then

$X = \sin\left(\dfrac{-\pi}{5}\right)$ (2)

but from (1) if $\arcsin X = \alpha \Rightarrow X = \sin\alpha$ (definition of arcsin)   (3)

so substituting (2) into (3) $\sin\left(\dfrac{-\pi}{5}\right) = \sin\alpha$

$\Rightarrow \left(\dfrac{-\pi}{5}\right) = \alpha$ (4)

Substituting (2) and (4) into (1) gives

$\arcsin\left(\sin\left(\dfrac{-\pi}{5}\right)\right) = \left(\dfrac{-\pi}{5}\right)$

2 a) To evaluate $\sin 120°$ let $\sin 120° = \sin 2A$ where $A = 60°$

   The double angle formula $\sin 2A \equiv 2\sin A\cos A$

   $\sin 120° = 2\sin 60°\cos 60°$ (written in terms of common angles means we can evaluate without using a calculator!)

   $\sin 120° = 2\times\dfrac{\sqrt{3}}{2}\times\dfrac{1}{2} = \dfrac{2\sqrt{3}}{4} = \dfrac{\sqrt{3}}{2}$

 b) To evaluate $\cos\pi^c$ let $\cos\pi^c = \cos 2A$ where $A = \dfrac{\pi}{2}^c$

   The double angle formula $\cos 2A \equiv \cos^2 A - \sin^2 A$

   $\Rightarrow \cos\pi^c = \cos^2\dfrac{\pi}{2} - \sin^2\dfrac{\pi}{2}$ (written in terms of common angles means we can evaluate without using a calculator!)

   $\Rightarrow \cos\pi^c = 0 - (1)^2 \Rightarrow \cos\pi^c = -1$

 c) To evaluate $\tan 90°$ let $\tan 90° = \tan 2A$ where $A = 45°$

   The double angle formula for $\tan 2A \equiv \dfrac{2\tan A}{1 - (\tan A)^2}$

   $\Rightarrow$ $\tan 90° = \dfrac{2\tan 45°}{1 - (\tan 45°)^2} = \dfrac{2\times 1}{1 - 1} = \infty$

Exponential functions

Exponential functions which are functions of the form $y = a^x$ were first introduced in Core 2. The constant a is a non-negative number called the **base** and x is the independent variable. The graphs of these functions take the same general shape.

Graphs of exponential functions

You saw in Core 2 that the general shape of the graph of an exponential function is similar to the graph of $y = 2^x$ shown below.

Fig. 3.1

Graph of $y = 2^x$.

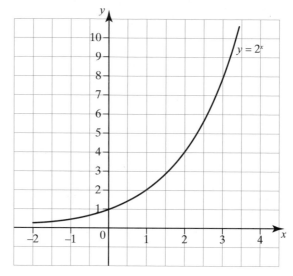

The graph passes through (0, 1) because $a^0 = 1$ for any number a. The value of y is always positive.

The gradient of the curve is always positive and gets larger (at an increasing rate) as positive values of x get larger. For negative values of x the graph approaches the x-axis but never reaches it. This can expressed as: as $x \rightarrow \infty$ then $y \rightarrow \infty$ and as $x \rightarrow -\infty$ then $y \rightarrow 0$

The gradient functions of exponential graphs are similar to the functions themselves as was explained in Core 2. For example, for $y = 2^x$ the gradient function is $\dfrac{dy}{dx} = 0.69 \times 2^x$ and for $y = 3^x$ the gradient function is $\dfrac{dy}{dx} = 0.1.1 \times 3^x$

Essential notes

The numbers 0.69 and 1.1 are given correct to 2 significant figures.

Example

Given that $f(x) = 2^x$ show that each of these statements is true:

a) $f(0) = 1$

b) $f(x + y) = f(x) + f(y)$

---

c) $\dfrac{f(x)}{f(y)} = f(x-y)$

d) $(f(x))^n = f(nx)$

Answer

a) $f(x) = 2^x \Rightarrow f(0) = 2^0 = 1$.

b) $f(x + y) = 2^{x+y}$

$\qquad = 2^x \times 2^y$

but $f(x) = 2^x$ and $f(y) = 2^y$ therefore:

$\quad f(x + y) = f(x) \times f(y)$

c) $\dfrac{f(x)}{f(y)} = \dfrac{2^x}{2^y}$

$\qquad = 2^{x-y}$

but $f(x) = 2^x$ therefore:

$\qquad \dfrac{f(x)}{f(y)} = f(x - y)$

d) $(f(x))^n = (2^x)^n$

$\qquad = (2)^{nx}$

but $f(x) = 2^x \Rightarrow (2)^{nx}$

$\qquad = f(nx)$, since $f(x) = 2^x$.

Essential notes

If the given function is $f(x) = 2^x$ then $f(y) = 2^y$.

Method notes

To solve this problem, use the rules for manipulating indices:

$a^m \times a^n = a^{m+n}$

$a^m \div a^n = a^{m-n}$

$(a^m)^n = a^{mn}$

The natural exponential function

Definition

We have looked at exponential functions of the form $y = a^x$. If $a = \text{e}$ where $\text{e} = 2.718281828459$ (to 12 d.p.) then $y = \text{e}^x$ is the 'natural' exponential function, often just called *the* exponential function. e is its base number and is *a* constant.

Essential notes

The number 'e' is similar to π in that it is an irrational number representing one of 'nature's numbers' occurring naturally in the real world.

Stop and think

a) Use your calculator to evaluate $\left(1 + \dfrac{1}{n}\right)^n$ with $n = 1$, $n = 2$, $n = 3$, $n = 4$, $n = 999$

b) Evaluate $\left(1 + \dfrac{1}{n}\right)^n$ n when $n = 2999$ and give the answer to 9 decimal places.

c) Given the definition $\text{e}^x = \lim\limits_{n \to \infty} \left(1 + \dfrac{x}{n}\right)^n$

by taking a suitable value for x and using your answer from b) find an approximation for the value of e to 9 decimal places.

Gradient function of the natural exponential function

If we investigate the gradients of the tangent to the curve

$y = e^x$ at different points on the curve we find that when:

$$x = 1 \text{ gradient function } \frac{dy}{dx} = 2.71828 = e^1$$

$$x = 2 \text{ gradient function } \frac{dy}{dx} = 7.38906 = e^2$$

$$x = 3 \text{ gradient function } \frac{dy}{dx} = 20.08553 = e^3$$

Essential notes

This is the only function for which $\dfrac{dy}{dx} = y$.

which shows a pattern emerging. At **any** point on a graph of the natural exponential function $y = e^x$ the gradient of the curve at that point is equal to the y-coordinate of the point chosen. More formally this is stated as:

if $y = e^x$ the gradient function $\dfrac{dy}{dx} = e^x$ for all values of x.

Graphs of natural exponential functions $y = e^{kx}$

Key points

The general form of the graph of the exponential function $y = e^{kx}$ depends on whether the constant k is positive or negative. Figure 3.2 shows the graphs of $y = e^{kx}$ for $k = 1$ and $k = 3$

When k is positive i.e. $k > 0$ the graphs show exponential growth because y increases as x increases.

Exam tips

Learn the shape of the graph of $y = e^x$.

Fig. 3.2
Graph of $y = e^{3x}$ transformed from graph of $y = e^x$.

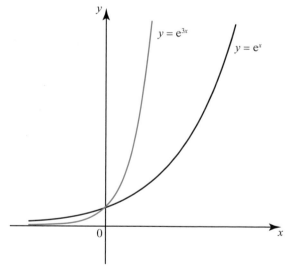

Essential notes

Recalling the transformations of graphs from Core 2 we see that the constant 3 is a stretch of $y = e^x$, factor $\frac{1}{3}$, parallel to the x-axis.

Figure 3.3 shows the graphs of $y = e^{kx}$ for $k = -1$ and $k = -2$

When k is negative i.e. $k < 0$ the graphs show exponential decay because y decreases as x increases.

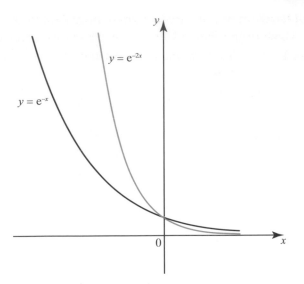

Fig 3.3
Graphs of $y = e^{-x}$ and $y = e^{-2x}$.

Essential notes

Recalling the transformations of graphs in Core 2 we see that the graph of $y = e^{-x}$ is a reflection of the graph of $y = e^x$ in the y-axis.

The graph of $y = e^{-2x}$ is a reflection of the graph of $y = e^x$ in the y-axis followed by a stretch factor $\frac{1}{2}$ parallel to the x-axis.

Example

On separate grids sketch a graph of $y = e^x$ and each of the following functions. In each case describe the geometrical transformation(s) which map(s) $y = e^x$ on to the given function.

a) $y = e^{2x-3}$

b) $y = e^{3x} - 5$

Answer

a) The geometric transformation from $y = e^x$ is a stretch, factor $\frac{1}{2}$, parallel to the x-axis followed by a translation through 1.5 units parallel to the x-axis.

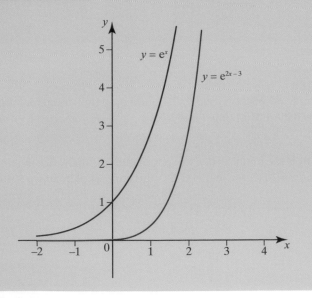

Fig. 3.4
Graphs of $y = e^x$ and $y = e^{2x-3}$.

Method notes

Rewrite $y = e^{2x-3}$ as

$y = e^{2(x-1.5)}$.

When deciding the order of the transformations consider first any multiplier of x. In this case $2x$ means we deal with the stretch factor of $\frac{1}{2}$ first, then $(x - 1.5)$ shows the translation of 1.5 units.

☞ Continued on the next page

b) The geometric transformation from $y = e^x$ is a stretch, factor $\frac{1}{3}$, parallel to the x-axis followed by a translation through -5 units parallel to the y-axis

Fig. 3.5
Graphs of $y = e^x$ and $y = e^{3x} - 5$

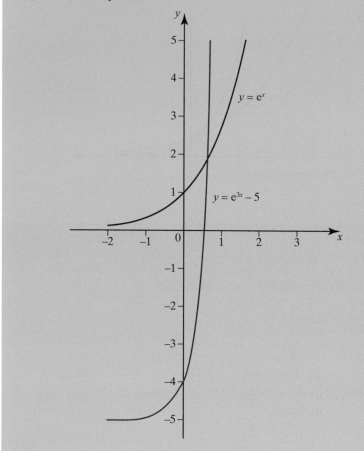

Method notes

As explained in the last example we deal with $3x$ first.

Example
In a simple population model, the number of individual members of the population, P, at time t is given by:

$$P = P_0 e^{(B-D)t}$$

where P_0 is the initial size of the population in 1000s, B is the number of births per 1000 per year and D is the number of deaths per 1000 per year.

a) What happens if :

   i) $B > D$

   ii) $B < D$

   iii) $B = D$?

b) If, in a certain population, $B = 0.25$, $D = 0.15$ and $P_0 = 10\,000$ estimate the size of the population in:

   i)   5 years

   ii)   10 years

   iii) 100 years.

c) Explain why, in the real world, your answer to b(iii) is probably unrealistic.

Answer

a) i) For $B > D$, the population will increase for ever.

   ii) For $B < D$, the population will decrease for ever.

   iii) For $B = D$, the population remain constant at P_0.

b) With $B = 0.25$, $D = 0.15$ and $P_0 = 10\,000$ the formula for the population becomes $P = 10000e^{0.1t}$

   i)   After 5 years $t = 5$ so $P = 10000e^{0.5} = 16487$

   ii)   After 10 years $t = 10$ so $P = 10000e^{1} = 27182$

   iii) After 100 years $t = 100$ so $P = 10000e^{10} = 2.2 \times 10^8$

c) The population is likely to be smaller than 2.2×10^8 for many reasons e.g. it is unlikely that there are the natural resources to support such a population.

Method notes

P_0 is a constant value throughout the question as it is the initial size of the population.

a) i) If $B > D$ then $B - D$ is positive so $e^{(B-D)t}$ is an increasing function therefore $P = P_0 e^{(B-D)t}$ increases.

   ii) and (iii) are answered using a similar method.

Essential notes

As explained in Core 2, the two statements $y = x^2$ and $x = \pm\sqrt{y}$ are called equivalent statements because the square root function 'undoes' squaring a number. The function which 'undoes' the exponential function is called the logarithmic function hence $x = \log_a y$ and $a^x = y$ are equivalent statements.

The logarithmic function

We saw in Core 2 that if $a^x = y$ then $x = \log_a y$ where a is called the base of the logarithm and of the exponential function, and that $x = \log_a y$ and $a^x = y$ are equivalent statements.

Key points about the logarithmic functions:

- $\log_a 1 = 0$ and $\log_a a = 1$
- $\log_{10} x$ is usually written as $\log x$
- The domain of the function $y = \log_{10} x$ is $\{x: x > 0\}$.

Essential notes

log x appears on most calculators as LOG.

Essential notes

It is important to commit these formulae to memory.

- The three laws of logarithms are
  - Product law: $\log_a(xy) = \log_a x + \log_a y$
  - Quotient law: $\log_a\left(\dfrac{x}{y}\right) = \log_a x - \log_a y$
  - Power law: $\log_a(x)^k = k\log_a x$

The natural logarithmic function

Essential notes

ln x appears on most calculators as LN.

The logarithm which has e as its base number is called the natural logarithm and is written as $\log_e x$ or ln x. This means that if $y = e^x$ the equivalent statement is $x = \ln y$.

The rules of manipulating natural logarithms are identical to those for logarithms to other bases. These are particularly important.

- $\ln 1 = 0$
- $\ln e = 1$
- The domain of the function $y = \ln x$ is $\{x: x > 0\}$.

The three laws of logarithms also apply to natural logarithms hence:

- $\ln(xy) = \ln x + \ln y$

Essential notes

It is important to commit these formulae to memory.

- $\ln\left(\dfrac{x}{y}\right) = \ln x - \ln y$

- $\ln(x)^k = k\ln x$

The natural logarithm function $y = \ln x$ is the inverse of the exponential function $y = e^x$.

In chapter 1 of this book we defined a one-one function f(x) as a function where each element of the domain set A mapped onto exactly one element in the range set B. The inverse function f^{-1}(x) is the mapping which maps elements in the range set B back to elements in the domain set A.

To show that the function $y = $ f(x) $= \ln x$ is the inverse of $y = e^x$ we will use the method explained in chapter 1.

Step 1: Make x the subject of $y = \ln x$ by writing the equivalent statement which is $e^y = x$.

Step 2: Interchange x and y in the last equation in step 1 which gives $e^x = y$.

Step 3: Write y as the inverse function so f^{-1}(x) $= e^x$. Therefore the natural logarithm function $y = \ln x$ is the inverse of the exponential function $y = e^x$.

The natural logarithmic function does not exist for negative values of x.

Graphs of natural logarithms

The graphs of $y = e^x$ and $y = \ln x$ are shown in the diagram below:

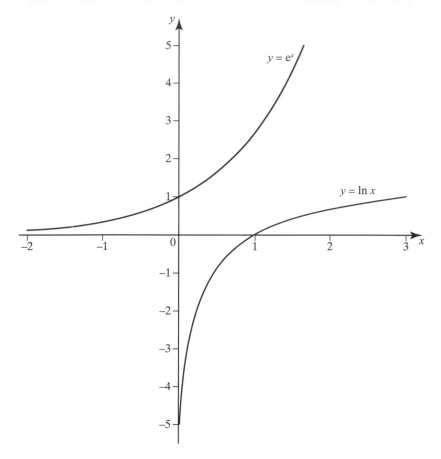

Fig. 3.6
The graphs of $y = e^x$ and $y = \ln x$.

Key points

As with all inverse functions, the two graphs are reflections of each other in the line $y = x$. For both as $x \to \infty$ then $y \to \infty$.

For the graph of $y = \ln x$:

- The graph passes through the point $(1, 0)$.

- As $x \to 0$ then $y \to -\infty$ and the negative y-axis is an asymptote to the curve

- $y = \ln x$ does not exist for $x < 0$

- The gradient of the curve is always positive and gets shallower as x gets larger.

Essential notes

The graphs of inverse functions were explained in chapter 1 of this book.

An asymptote is a line along which a curve approaches ∞.

Method notes

If you try to evaluate ln of any negative number your calculator shows 'error'.

Example

Sketch the graphs of

a) $y = \ln(4 - x)$

b) $y = 2 + \ln(3x)$

Answer

a) $y = \ln(4 - x)$

Fig. 3.7
Graph of $y = \ln(4 - x)$.

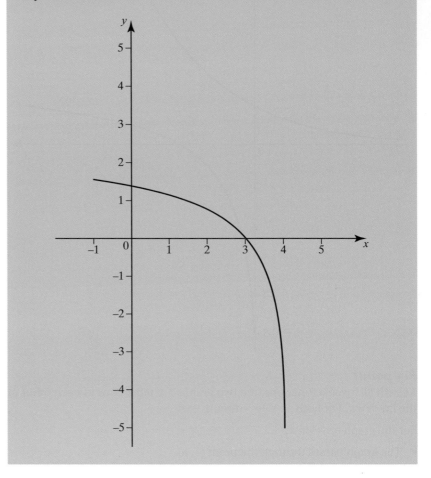

Method notes

As the natural logarithmic function does not exist for negative values of x, in this case if $(4 - x) < 0$ then $x > 4$ which means that y has no values to the right of the line $x = 4$

As $x \to 4$ then $y \to \ln 0$ so $y \to -\infty$.

For $x = 0$, $y = \ln 4 = 1.386$ (3 d.p.).

For $y = 0$, $(4 - x) = 1 \Rightarrow x = 3$

b) $y = 2 + \ln(3x)$

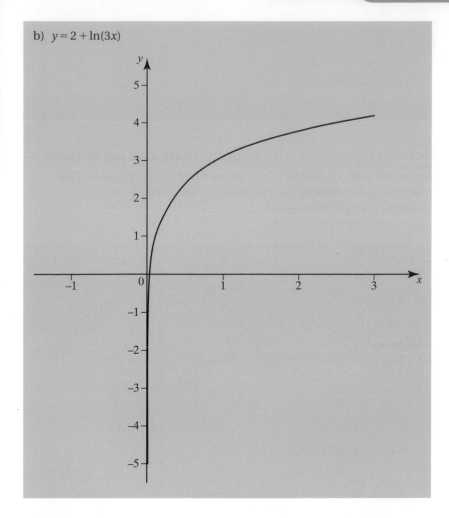

Fig. 3.8
Graph of $y = 2 + \ln(3x)$.

Method notes

As $x \to 0$ then $y \to -\infty$

As $x \to \infty$ then $y \to \infty$

For $y = 0$ $2 + \ln(3x) = 0$

$\Rightarrow \ln(3x) = -2$

leading to the equivalent statement

$3x = e^{-2}$

$\Rightarrow x = \frac{1}{3}e^{-2}$

$\quad = 0.045$ (3 d.p.).

Example

The functions f and g are defined by

$f : x \mapsto 2x + \ln x, \, x > 0, \, x \in \Re$

$g : x \mapsto e^{x^2}, \, x \in \Re$

a) Write down the range of g.

b) Define the composite function fg.

c) Write down the range of fg

Answer

a) The range of $y = g(x)$ is $\{y : 1 < y < \infty\}$

b) $fg(x) = f(e^{x^2}) = 2e^{x^2} + \ln(e^{x^2})$

Method notes

The range of a function was explained in chapter 1. $x \in \Re$ means that x is any real number.

a) We know that $g(x) = e^{x^2}$ and the smallest value of $e^{\text{positive number}} = 1$ but the largest value is ∞.

☞ Continued on the next page

3 The exponential and logarithmic functions

Method notes

b) Remember that in the expression for fg(x) the domain of f is the range of g. The smallest value of fg(x) is found by using the answer from a) i.e. substituting $x = 1$ in the formula for fg. For the largest value substitute $x = \infty$.

Using law 3 of logarithms $= 2e^{x^2} + x^2 \ln e$

But $\ln e = 1$ so fg(x) $= 2e^{x^2} + x^2$

Therefore fg $: x \mapsto 2e^{x^2} + x^2$

c) The range of $y = $ fg(x) is $\{y: (2e + 1) < y < \infty\}$.

Solving equations with exponentials and logarithms

Logarithm and exponential functions are inverses of each other. This relationship can often be used to solve a problem where one of the functions needs to be reversed.

Example
Solve the equations:

a) $e^{3x-2} = 5$

b) $3 \ln x + 2 = 7$

Answer
a) **Step 1:** To solve $e^{3x-2} = 5$ write the equivalent statement

$$3x - 2 = \ln 5 \qquad (1)$$

Step 2: From equation (1) $3x = \ln 5 + 2$

$$\Rightarrow \qquad x = \frac{1}{3}(\ln 5 + 2)$$

b) **Step 1:** To solve $3 \ln x + 2 = 7$

$$\text{we need to solve } 3 \ln x = 5 \qquad (1)$$

Step 2: Divide both sides of equation (1) by 3

$$\Rightarrow \qquad \ln x = \frac{5}{3} \qquad (2)$$

Step 3: Write the equivalent statement for equation (2)

$$\Rightarrow \qquad x = e^{\frac{5}{3}}$$

Method notes

By writing the equivalent statement it means you have reversed the form in which the function is written.

Example
Solve the equation $e^{4x} - e^{2x} - 6 = 0$

Answer
Step 1: To solve $e^{4x} - e^{2x} - 6 = 0 \qquad (1)$

let $u = e^{2x}$ so that using rules of indices:

$$e^{4x} = (e^{2x})^2 = u^2 \qquad (2)$$

Step 2: Rewrite equation (1) in terms of u using equation (2), then factorise:

$$u^2 - u - 6 = (u - 3)(u + 2) = 0 \qquad (3)$$

Step 3: Solve equation 3:

$u = 3$ or $u = -2$

Step 4: Substitute the values of u from step 3 into the original statement for u in step 1:

If $u = 3$, then $e^{2x} = 3$ and the equivalent statement is:

$2x = \ln(3)$

$\Rightarrow \qquad x = \frac{1}{2}\ln 3 = 0.5493$

If $u = -2$, then $e^{2x} = -2$ which is not possible, since $e^{2x} > 0$ for any real value of x

So the only solution is $x = \dfrac{1}{2}\ln 3 = 0.5493$ (4 d.p.)

Example
A cup of coffee, at temperature $T^{o}C$ cools down according to the formula:

$$T = 70e^{-kt} + 20$$

where t is the time in seconds after the cup of coffee was made.

a) What is the initial temperature of the coffee?

b) The coffee cools by $20^{o}C$ after the first minute. Find the value of k.

c) How long will it take for the temperature to drop to $50^{o}C$?

Answer
a) The initial temperature is found by substituting $t = 0$ in the formula for T.

$T = 70e^{-k \times 0} + 20 = 70 + 20 = 90$

The initial temperature of the coffee is $90^{o}C$.

b) When $t = 60$, $T = 90 - 20 = 70$. Substituting these values into the formula for T gives:

$70 = 70e^{-k \times 60} + 20$

$\Rightarrow \qquad 70e^{-60k} = 50$

$\Rightarrow \qquad e^{-60k} = \dfrac{50}{70} = \dfrac{5}{7}$ so the equivalent statement is:

$-60k = \ln\left(\dfrac{5}{7}\right)$

$k = -\left(\dfrac{1}{60}\right)\ln\left(\dfrac{5}{7}\right) = 0.0056$

Method notes

a) Initial means when $t = 0$
Remember that $e^0 = 1$

b) Care must be taken in using the original units given for the variables throughout the question. In this case t was given in seconds so $t = 60$ after 1 minute. If the temperature **cools** by 20° this means the temperature is 20° **less** than 90°.

☞ Continued on the next page

c) We need to find the value of t when $T = 50$ so substituting $T = 50$ into the formula for T:

$$\Rightarrow \qquad 50 = 70e^{-0.0056t} + 20$$

$$\Rightarrow \qquad e^{-0.0056t} = \frac{30}{70} = \frac{3}{7}$$

The equivalent statement is:

$$-0.0056t = \ln\frac{30}{70} = -\tfrac{1}{0.0056}\ln\frac{3}{7} = 151$$

Therefore the temperature cools to 50°C after 151 seconds or 2 min 31s.

Stop and think answers

a) When $n = 1$: $\left(1 + \dfrac{1}{n}\right)^n = (1 + 1)^1 = 2.00000$

$n = 2$: $\left(1 + \dfrac{1}{n}\right)^n = \left(1 + \dfrac{1}{2}\right)^2 = (1.5)^2 = 2.25000$

$n = 3$: $\left(1 + \dfrac{1}{n}\right)^n = \left(1 + \dfrac{1}{3}\right)^3 = (1.333333)^3 = 2.37037....$

$n = 4$: $\left(1 + \dfrac{1}{n}\right)^n = \left(1 + \dfrac{1}{4}\right)^4 = (1.25)^4 = 2.44140....$

$n = 999$: $\left(1 + \dfrac{1}{n}\right)^n = \left(1 + \dfrac{1}{999}\right)^{999} = 2.71692$

b) When $n = 2999$: $\left(1 + \dfrac{1}{n}\right)^n = \left(1 + \dfrac{1}{2999}\right)^{2999} = 2.717828769$ (9 d.p)

c) From the definition $e^x = \lim_{n\to\infty}\left(1 + \dfrac{x}{n}\right)^n$, if $x = 1$ then:

$e = \lim_{n\to\infty}\left(1 + \dfrac{x}{n}\right)^n$ and from (b) we can see that if $n = 2999$ (which is large value of n) then:

$\left(1 + \dfrac{1}{n}\right)^n = 2.717828769$

Using $n = 10^{10}$, $n = 10^{11}$ and $n = 10^{12}$, we get the same answer for $\left(1 + \dfrac{1}{n}\right)^n$ to 9 decimal places of 2.71828183

So if n gets larger and approaches ∞ then a reasonable approximation for e = 2.717828183 (to 9 d.p.).

Rules of differentiation for composite functions

The rule for differentiating powers of x (or any other variable) is:

$$\text{if } y = x^n \text{ then } \frac{dy}{dx} = nx^{n-1} \text{ where } n \text{ is a constant.}$$

Essential notes

In Core 2 you used this rule when exploring increasing and decreasing functions and finding stationary points.

This was introduced in Core 1.

We now look at differentiating functions such as $y = (x^2 + 1)^3$ or $y = (x^4 + 1)^5$. These are examples of **composite functions**. Each is a **function of a function**.

This means that if we consider the function $y = (x^2 + 1)^3$ and let $u = (x^2 + 1)$ then $y = u^3$ so y is a function of u which in turn is a function of x. So y is a function of a function of x.

Example
Differentiate $y = (x^2 + 1)^3$ with respect to x.

Answer
Step 1: Rewrite $y = (x^2 + 1)^3$

$$= (x^2 + 1)(x^2 + 1)^2$$
$$= (x^2 + 1)(x^4 + 2x^2 + 1)$$
$$= x^6 + 3x^4 + 3x^2 + 1$$

Step 2: Apply the rule of differentiation to each term

$$\Rightarrow \frac{dy}{dx} = 6x^5 + 12x^3 + 6x$$

This answer can also be factorised to give

$$\frac{dy}{dx} = 6x(x^4 + 2x^2 + 1) = 6x(x^2 + 1)^2$$

and this can also be written as:

$$\frac{dy}{dx} = 3 \times 2x \times (x^2 + 1)^2$$

Therefore if $y = (x^2 + 1)^3$ then $\dfrac{dy}{dx} = 3 \times 2x \times (x^2 + 1)^2$

From this example you can see that in the answer:

- the power of 3 from $(x^2 + 1)^3$ becomes a multiplier
- the power of the bracket $(x^2 + 1)^3$ was 3 and it is now reduced by 1 to become 2
- the derivative of the function $x^2 + 1$ inside the bracket is $2x$, and is also a multiplier.

If we now start with the function $y = (x^4 + 1)^5$ and apply the observations listed above to find $\dfrac{dy}{dx}$ we would see in the answer:

- the power of 5 from $(x^4 + 1)^5$ becomes a multiplier
- the power of the bracket $(x^4 + 1)^5$ was 5 and it is reduced by 1 to become 4
- the derivative of the function '$x^4 + 1$' inside the bracket is $4x^3$, and is also a multiplier.

Therefore if $y = (x^4 + 1)^5$ then $\dfrac{dy}{dx} = 5 \times 4x^3 \times (x^4 + 1)^4$

$\Rightarrow$ $\qquad\qquad\qquad\qquad\qquad \dfrac{dy}{dx} = 20x^3 (x^4 + 1)^4$

From our previous working given:

$y = (x^2 + 1)^3$ then $\dfrac{dy}{dx} = 3 \times 2x \times (x^2 + 1)^2$ $\qquad\qquad\qquad$ (1)

If $y = u^3$ then $\dfrac{dy}{du} = 3u^2$

and as $u = (x^2 + 1)$ then $\dfrac{du}{dx} = 2x$

If we now rewrite (1) in terms of u we have

$\dfrac{dy}{dx} = 3 \times \dfrac{du}{dx} \times u^2$

$\Rightarrow \dfrac{dy}{dx} = 3u^2 \times \dfrac{du}{dx}$

$\Rightarrow \dfrac{dy}{dx} = \dfrac{dy}{du} \times \dfrac{du}{dx}$

This is known as the **chain rule** and we use it to differentiate composite functions (also known as function of a function rule).

Example
Use the chain rule to differentiate the function $y = (x^4 + 1)^5$

Answer
Step 1: Let $u = x^4 + 1$ and write y in terms of $u \Rightarrow y = u^5$

Step 2: As $y = u^5$ find $\dfrac{dy}{du} \Rightarrow \dfrac{dy}{du} = 5u^4$

Step 3: As $u = x^4 + 1$ find $\dfrac{du}{dx} \Rightarrow \dfrac{du}{dx} = 4x^3$

Step 4: State the chain rule which is $\dfrac{dy}{dx} = \dfrac{dy}{du} \times \dfrac{du}{dx}$ and substitute into this the results from steps 2 and 3:

$\Rightarrow \dfrac{dy}{dx} = 5u^4 \times 4x^3$

Step 5: Rewrite the answer from step 4 in terms of x

$\Rightarrow \dfrac{dy}{dx} = 5(x^4 + 1)^4 \times 4x^3 = 20x^3 (x^4 + 1)^4$

This is the answer we obtained earlier using another method!

Essential notes

Function notation was explained in Core 2.

f(u) means f is a function of the variable u.

g(x) means g is a function of the variable x.

The chain rule for differentiation

- If $y = [f(x)]^n$ then $\dfrac{dy}{dx} = n[f(x)]^{n-1}f'(x)$

or

- If $y = f(u)$ with $u = g(x)$ then $\dfrac{dy}{dx} = \dfrac{dy}{du} \times \dfrac{du}{dx}$

Example

Differentiate $y = (3x^2 + 5)^2$ with respect to x

a) by expanding the brackets

b) using the chain rule.

Show that the two results are the same.

Answer

a) $y = (3x^2 + 5)^2 = 9x^4 + 30x^2 + 25$

$\dfrac{dy}{dx} = 36x^3 + 60x = 12x(3x^2 + 5)$

b) For $y = (3x^2 + 5)^2$

let $u = 3x^2 + 5 \Rightarrow \dfrac{du}{dx} = 6x$

and as $y = u^2 \Rightarrow \dfrac{dy}{du} = 2u$

The chain rule states that: $\dfrac{dy}{dx} = \dfrac{dy}{du} \times \dfrac{du}{dx}$

$\Rightarrow \dfrac{dy}{dx} = 2u \times 6x$

which written in terms of $x \Rightarrow \dfrac{dy}{dx} = 12x(3x^2 + 5)$

This is the same as the answer in a).

Method notes

You will find it helpful to introduce a new variable, say u, when using the chain rule. This variable is then the 'basic' function (often the function inside a bracket before it is raised to a power). You can choose any letter for this new variable.

Method notes

In this example choose the 'basic' function to be $(5x^2 + 3x)$.

Example

Differentiate $y = \sqrt{(5x^2 + 3x)}$ with respect to x using the chain rule.

Answer

For $y = \sqrt{(5x^2 + 3x)} = (5x^2 + 3x)^{\frac{1}{2}}$

let $u = 5x^2 + 3x$

$\Rightarrow \dfrac{du}{dx} = 10x + 3$

and as $y = \sqrt{u} = u^{\frac{1}{2}}$

$\Rightarrow \dfrac{dy}{du} = \dfrac{1}{2}(u)^{-\frac{1}{2}}$

$\Rightarrow \quad = \dfrac{1}{2\sqrt{u}}$

The chain rule states that: $\dfrac{dy}{dx} = \dfrac{dy}{du} \times \dfrac{du}{dx}$

$$\Rightarrow \frac{dy}{dx} = \frac{1}{2\sqrt{u}} \times (10x + 3)$$

which written in terms of x:

$$\Rightarrow \frac{dy}{dx} = \frac{10x + 3}{2\sqrt{(5x^2 + 3x)}}$$

Example

Find the equation of the tangent to the curve with equation $y = (5 - 2x)^3$ at the point $(1, 27)$.

Answer

The tangent to a curve is a straight line and its equation can be found by using the formula $\dfrac{y - y_1}{x - x_1} = m$ where m is the gradient of the line.

Step 1: For $y = (5 - 2x)^3$ let $u = 5 - 2x \Rightarrow \dfrac{du}{dx} = -2$

Step 2: If $u = 5 - 2x \Rightarrow y = u^3 \Rightarrow \dfrac{dy}{du} = 3u^2$

Step 3: State the chain rule and substitute the results from steps 1 and 2:

$$\frac{dy}{dx} = \frac{dy}{du} \times \frac{du}{dx}$$

$$\Rightarrow \frac{dy}{dx} = 3u^2 \times (-2)$$

$$= -6u^2$$

Step 4: Rewrite the answer from step 3 in terms of x

$$\Rightarrow \frac{dy}{dx} = -6(5 - 2x)^2$$

which gives the gradient of the tangent to the curve at any point.

Step 5: At the point $(1, 27)$, $x = 1$ so using step 4, the slope of the tangent is:

$$\frac{dy}{dx} = -6(5 - 2)^2 = -54$$

Step 6: Use the formula $\dfrac{y - y_1}{x - x_1} = m$ with $m = -54$ and $x_1 = 1$ and $y_1 = 27$

to find the equation of the tangent:

$$\frac{y - 27}{x - 1} = -54$$

$$\Rightarrow \quad y = -54x + 81$$

Method notes

The equation of a tangent to a curve was covered in Core 2. The gradient of the tangent to a curve at any point is $\dfrac{dy}{dx}$.

Relationship between $\dfrac{dy}{dx}$ and $\dfrac{dx}{dy}$

In all the examples of differentiation that you have seen so far you have been given y in terms of x and you have been asked to find $\dfrac{dy}{dx}$. The next examples illustrate how to find $\dfrac{dx}{dy}$.

Example

Given $y = x^2$ find in terms of x:

a) $\dfrac{dy}{dx}$

b) $\dfrac{dx}{dy}$.

c) Using your answers from a) and b) find the relationship between $\dfrac{dy}{dx}$ and $\dfrac{dx}{dy}$.

Answer

If $y = x^2$ then $\dfrac{dy}{dx} = 2x$

If $y = x^2$

$\Rightarrow y^{\frac{1}{2}} = x$

or

$x = y^{\frac{1}{2}}$

$\Rightarrow \dfrac{dx}{dy} = \dfrac{1}{2}y^{-\frac{1}{2}}$ using the rule of differentiation

so substituting $y = x^2$

$\Rightarrow \dfrac{dx}{dy} = \dfrac{1}{2}(x^2)^{-\frac{1}{2}} = \dfrac{1}{2}\left(\dfrac{1}{\sqrt{x^2}}\right)$ using rules of indices

$= \dfrac{1}{2}\left(\dfrac{1}{x}\right)$

So $\dfrac{dx}{dy} = \dfrac{1}{2x}$

from a) $\dfrac{dy}{dx} = 2x$ and from b) $\dfrac{dx}{dy} = \dfrac{1}{2x}$

$\Rightarrow \dfrac{dy}{dx} = \dfrac{1}{\left(\dfrac{dx}{dy}\right)}$

Method notes

More generally:

from the chain rule

$\dfrac{dy}{dx} \times \dfrac{dx}{dy} = \dfrac{dy}{dy} = 1$

$\Rightarrow \dfrac{dy}{dx} = \dfrac{1}{\left(\dfrac{dx}{dy}\right)}$

Exam tips

You must learn this result as it is not in the examination booklet.

Example

Use the result $\dfrac{dy}{dx} = \dfrac{1}{\left(\dfrac{dx}{dy}\right)}$ to find the value of $\dfrac{dy}{dx}$ for the curve with

equation
$x = 2y^3 + y^2 - 4y + 5$

Answer

Differentiating $x = 2y^3 + y^2 - 4y + 5$ with respect to y

$\Rightarrow \dfrac{dx}{dy} = 6y^2 + 2y - 4$

but

$\dfrac{dy}{dx} = \dfrac{1}{\left(\dfrac{dx}{dy}\right)}$

$\Rightarrow \dfrac{dy}{dx} = \dfrac{1}{6y^2 + 2y - 4}$

The product rule for differentiation

$y = x^3(x+1)^2$ is an example of a **product** of two functions of x: the first function is x^3 and the second function is $(x+1)^2$.

To differentiate this product, first expand the brackets:

$y = x^3(x+1)^2 = x^3(x^2 + 2x + 1) = x^5 + 2x^4 + x^3$

then each term can be differentiated with respect to x so

$\dfrac{dy}{dx} = 5x^4 + 8x^3 + 3x^2$

Another way of differentiating products of functions without the need to expand brackets is to use the **product rule**.

Proving the product rule

Take two functions of x, p(x) and q(x). Let $u = $ p(x) and

$v = $ q(x).

In Core 1 we introduced a formal definition of the gradient function as:

f$'(x)$ is the limit of $\dfrac{f(x+h) - f(x)}{h}$ as $h \to 0$

Consider the product $y = uv$ or $y = $ f$(x) = $ p(x)q(x).

f$(x+h) - $ f$(x) = $ p$(x+h)$q$(x+h) - $ p(x)q(x)

$\qquad = $ p$(x+h)$q$(x+h) - $ p(x)q$(x+h) + $ p(x)q$(x+h) - $ p(x)q(x)

$\qquad = [$p$(x+h) - $ p$(x)]$q$(x+h) + $ p$(x)[$q$(x+h) - $ q$(x)]$

Essential notes

Function notation was explained in Core 2.

$u = $ p(x) means u is a function of x. If we differentiate u with respect to x we write

$\dfrac{du}{dx} = $ p$'(x)$.

$v = $ q(x) means v is another function of x. If we differentiate v with respect to x we write $\dfrac{dv}{dx} = $ q$'(x)$

Method notes

The two extra terms

$-$p(x)q$(x+h) + $ p(x)q$(x+h)$

were introduced to rewrite the expression

f$(x+h) - $ f(x).

4 Differentiation

Exam tips

You do not need to know how to prove the product rule but you must learn the result and how to apply it as it will not be in your formulae booklet.

Dividing both sides by h

$$\frac{f(x+h)-f(x)}{h} = \frac{[p(x+h)-p(x)]q(x+h)+p(x)[q(x+h)-q(x)]}{h}$$

$$= \left[\frac{p(x+h)-p(x)}{h}\right]q(x+h)+p(x)\left[\frac{q(x+h)-q(x)}{h}\right]$$

Taking the limit as $h \to 0$ gives:

$$f'(x) = p'(x)\,q(x) + p(x)\,q'(x)$$

or $\dfrac{dy}{dx} = \left(\dfrac{du}{dx} \times v\right) + \left(u \times \dfrac{dv}{dx}\right)$

This is called the **product rule** for differentiation. We will now apply this rule to our original example.

Example

Find $\dfrac{dy}{dx}$ for the function $y = x^3(x+1)^2$

Answer

You know that this is a product of two functions of x.

Step 1: Let $u = x^3 \Rightarrow \dfrac{du}{dx} = 3x^2$

Step 2: Let $v = (x+1)^2$ which is a function of a function of x. We need to use the chain rule to find $\dfrac{dv}{dx}$.

$$\Rightarrow \frac{dv}{dx} = 2(x+1)^1 \times 1$$
$$= 2(x+1)$$

Step 3: Apply the product rule :

$$\frac{dy}{dx} = \frac{du}{dx} \times v + u \times \frac{dv}{dx} = (3x^2 \times (x+1)^2) + (x^3 \times 2(x+1))$$

$$= (x+1)(3x^2(x+1)) + (2x^3)$$
$$= x^2(x+1)[3(x+1)+2x]$$
$$= x^2(x+1)(5x+3)$$
$$= 5x^4 + 8x^3 + 3x^2$$

which is the answer we obtained previously.

Method notes

To differentiate $\sqrt{5x-1}$, write $\sqrt{5x-1} = (5x-1)^{\frac{1}{2}}$

Then using the chain rule,

$$\frac{dv}{dx} = \frac{1}{2} \times 5(5x-1)^{-\frac{1}{2}}$$

This is how $\dfrac{dv}{dx} = \dfrac{5}{2\sqrt{5x-1}}$ is obtained.

Example

Find $\dfrac{dy}{dx}$ for the function $y = (x^2+1)\sqrt{5x-1}$

Answer

In this case $u = x^2+1$ and $v = \sqrt{5x-1}$

$$\frac{du}{dx} = 2x \text{ and } \frac{dv}{dx} = \frac{5}{2\sqrt{5x-1}}$$

Applying the product rule gives:

$$\frac{dy}{dx} = \left(\frac{du}{dx} \times v\right) + \left(u \times \frac{dv}{dx}\right)$$

$$= 2x\sqrt{5x - 1} + (x^2 + 1)\frac{5}{2\sqrt{5x - 1}}$$

$$= \frac{4x(5x - 1) + 5(x^2 + 1)}{2\sqrt{5x - 1}}$$

$$\frac{dy}{dx} = \frac{25x^2 - 4x + 5}{2\sqrt{5x - 1}}$$

The quotient rule for differentiation

$y = f(x) = \dfrac{x + 5}{x + 2}$ is an example of the **quotient** of two functions of x.

We can write $y = \dfrac{x + 5}{x + 2} = (x + 5)(x + 2)^{-1}$ and then use the product rule

to differentiate y, showing that:

$u = (x + 5)$ so $\dfrac{du}{dx} = 1$

and $v = (x + 2)^{-1}$ so $\dfrac{dv}{dx} = -1(x + 2)^{-2}$

$$\frac{dy}{dx} = 1(x + 2)^{-1} + (x + 5) \times [-1(x + 2)^{-2}]$$

$$= \frac{1}{(x + 2)} - \frac{(x + 5)}{(x + 2)^2}$$

$$\frac{dy}{dx} = \frac{-3}{(x + 2)^2}$$

Proving the quotient rule

We will now develop another way of differentiating the quotient of two functions.

If $y = \dfrac{(x + 5)}{(x + 2)}$, let u be the function which is in the 'numerator' position of the fraction: in this case $u = (x + 5)$ and let v be the function which is in the denominator position of the fraction: in this case $v = (x + 2)$.

This means we can write the quotient as $y = \dfrac{u}{v}$

which can also be written as $y = u \times v^{-1}$ and then we can use the product rule to differentiate:

$$\frac{dy}{dx} = \left(\frac{du}{dx} \times v^{-1}\right) + u \times \left(-1v^{-2} \times \frac{dv}{dx}\right)$$

$$= \frac{du}{dx} \times \frac{1}{v} - \frac{u}{v^2} \times \frac{dv}{dx}$$

$$= \frac{du}{dx} \times \frac{1v}{v^2} - \frac{u}{v^2} \times \frac{dv}{dx}$$

Essential notes

A quotient is formed when two quantities are divided: it may be easier to think of a quotient as a fraction = $\dfrac{\text{numerator}}{\text{denominator}}$.

$$= \frac{1}{v^2}\left(v\frac{du}{dx} - u\frac{dv}{dx}\right)$$

$$\frac{dy}{dx} = \frac{v\frac{du}{dx} - u\frac{dv}{dx}}{v^2}$$

This is the quotient rule for differentiation.

Example

Find $\frac{dy}{dx}$ for the function $y = \frac{x + 5}{x + 2}$ using the quotient rule.

Answer

Step 1: Let $u = x + 5 \Rightarrow \frac{dy}{dx} = 1$ (u must be the 'numerator' function)

Step 2: Let $v = x + 2 \Rightarrow \frac{dv}{dx} = 1$ (v must be the 'denominator' function)

Step 3: Apply the quotient rule:

$$\frac{dy}{dx} = \frac{v\frac{du}{dx} - u\frac{dv}{dx}}{v^2}$$

$$= \frac{(x + 2) \times 1 - (x + 5) \times 1}{(x + 2)^2}$$

$$\frac{dy}{dx} = \frac{-3}{(x + 2)^2}$$

Example

a) Find and classify the stationary points of the curve with equation

$$y = f(x) = \frac{x}{x^2 + 4}$$

b) Sketch a graph of $y = \frac{x}{x^2 + 4}$

Answer

a) Given $y = \frac{x}{x^2 + 4}$ let $u = x$ (u must be the numerator

function) therefore $\frac{du}{dx} = 1$ and let $v = x^2 + 4$ (v must be the

denominator function) therefore $\frac{dv}{dx} = 2x$

The quotient rule gives

$$\frac{dy}{dx} = f'(x) = \frac{v\frac{du}{dx} - u\frac{dv}{dx}}{v^2} = \frac{(x^2 + 4) \times 1 - x \times 2x}{(x^2 + 4)^2}$$

$$\frac{dy}{dx} = f'(x) = \frac{4 - x^2}{(x^2 + 4)^2} = \frac{(2 - x)(2 + x)}{(x^2 + 4)^2}$$

Stationary points occur when $\frac{dy}{dx} = 0$

$$\Rightarrow \frac{(2 - x)(2 + x)}{(x^2 + 4)^2} = 0$$

$\Rightarrow (2-x)(2+x)=0$

$\Rightarrow x=2$ or $x=-2$

If $x=2$ in $y = f(x) = \dfrac{x}{x^2 + 4} \Rightarrow y = \dfrac{1}{4}$

If $x=-2$ in $y = f(x) = \dfrac{x}{x^2 + 4} \Rightarrow y = \dfrac{-1}{4}$

Therefore there are two stationary points: $\left(2, \dfrac{1}{4}\right)$ and $\left(-2, -\dfrac{1}{4}\right)$

To classify the stationary points, investigate the gradient either side of each stationary point:

close to $\left(2, \dfrac{1}{4}\right)$: take $x=1.5$ and $x=2.5$

$\Rightarrow f'(1.5) = \dfrac{1.75}{39.06}$ and $f'(2.5) = \dfrac{-2.25}{105.06}$

so the gradient changes from positive to negative and the point $\left(2, \dfrac{1}{4}\right)$ is a local maximum.

close to $\left(-2, -\dfrac{1}{4}\right)$: take $x=-2.5$ and $x=-1.5$

$\Rightarrow f'(-2.5) = \dfrac{-2.25}{105.06}$ and $f'(-1.5) = \dfrac{1.75}{39.06}$

so the gradient changes from negative to positive and the point $\left(-2, -\dfrac{1}{4}\right)$ is a local minimum.

b) To sketch the graph of $y = \dfrac{x}{x^2 + 4}$

let $x=0$ then $y=0$ and the graph passes through the origin

if $y \to 0$ then $x \to \pm\infty$

We know that $\left(2, \dfrac{1}{4}\right)$ is a maximum turning point and $\left(-2, -\dfrac{1}{4}\right)$ is a minimum turning point.

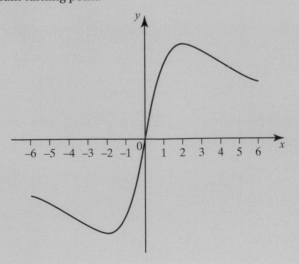

Essential notes

$(4-x^2)$ is the difference of two squares $\Rightarrow$

$(4-x^2) = (2+x)(2-x)$

Method notes

Classification of stationary points was covered in Core 2. It can be done using the second derivative but in examples such as this to differentiate $f'(x)$ is very time consuming. It is much quicker to use the method of investigating the gradient of the curve either side of the stationary point.

Fig. 4.1

The graph of $y = \dfrac{x}{x^2 + 4}$

4 Differentiation

Essential notes

This is the only function for which this statement is true

i.e. $\dfrac{dy}{dx} = y$

Differentiation of exponential functions

The natural exponential function $y = e^x$ was introduced in Chapter 3 and one of its important properties is that the gradient $\dfrac{dy}{dx}$ of the natural exponential function $y = e^x$ is identical to the function itself so $\dfrac{dy}{dx} = e^x$.

Proof

This result can be proved using the formal definition of the gradient function:

$f'(x)$ is the limit of $\dfrac{f(x+h) - f(x)}{h}$ as $h \to 0$

If $f(x) = e^x$ this means that $f(x + h) = e^{x+h}$ therefore:

$$\frac{f(x+h) - f(x)}{h} = \frac{e^{x+h} - e^x}{h} = e^x\left(\frac{e^h - 1}{h}\right)$$

The following table shows what happens to $\dfrac{e^h - 1}{h}$ as h gets smaller.

Method notes

The rule of indices gives:

$(e^x) \times (e^h) = e^{x+h}$

h	0.1	0.01	0.001	0.0001	0.00001
$\dfrac{e^h - 1}{h}$	1.05171	1.005017	1.000500	1.000050	1.000005

The table suggests that the expression $\dfrac{e^h - 1}{h}$ tends towards the value 1 as h tends to 0

This means that $e^x\left(\dfrac{e^h - 1}{h}\right) \to e^x$ as $h \to 0$ so the result is proved.

If $y = e^x$ then $\dfrac{dy}{dx} = e^x$

We now extend this to the differentiation of other exponential functions.

Example

Find $\dfrac{dy}{dx}$ if $y = e^{4x}$.

Answer

Step 1: If $y = e^{4x}$ this can be written as $y = (e^x)^4$ using rules of indices.

This is then a function of a function of x so we use the chain rule to find $\dfrac{dy}{dx}$.

Step 2: Let $u = e^x \Rightarrow \dfrac{du}{dx} = e^x$

and $y = (u)^4 \Rightarrow \dfrac{dy}{du} = 4u^3$

Step 3: Apply the chain rule which states $\dfrac{dy}{dx} = \dfrac{dy}{du} \times \dfrac{du}{dx}$

$$\Rightarrow \frac{dy}{dx} = 4u^3(e^x)$$

Step 4: Write in terms of x

$$\Rightarrow \frac{dy}{dx} = 4(e^x)^3(e^x) = 4\,(e^x)^4 = 4e^{4x} \text{ (using rules of indices)}$$

$$\text{Therefore if } y = e^{4x} \text{ then } \frac{dy}{dx} = 4e^{4x}$$

In general terms:
if $y = e^{ax}$ where a is a constant then $\dfrac{dy}{dx} = ae^{ax}$

Exam tips

You must learn this formula as it is not given in the examination booklet.

a is a constant and may be represented by any letter. k is used in the formula book.

Example

Differentiate the following with respect to x:

a) $y = 5e^{3x}$ b) $y = x\,e^{2x}$

c) $y = e^{x^2 + x - 3}$ d) $y = \dfrac{e^{4x}}{x^2}$

Answer

a) $y = 5e^{3x} \Rightarrow \dfrac{dy}{dx} = 5 \times 3e^{3x}$

$$= 15e^{3x}$$

b) $y = x\,e^{2x}$ is a product of two functions of x:

$$u = x \Rightarrow \frac{du}{dx} = 1$$

$$v = e^{2x} \Rightarrow \frac{dv}{dx} = 2e^{2x}$$

The product rule states:

$$\frac{dy}{dx} = \left(\frac{du}{dx} \times v \right) + \left(u \times \frac{dv}{dx} \right)$$

$$\Rightarrow \frac{dy}{dx} = (1 \times e^{2x}) + (x \times 2e^{2x}) = (1 + 2x)e^{2x}$$

c) $y = e^{x^2 + x - 3}$

Let $u = x^2 + x - 3$ then $u = f(x)$ so u is a function of x and $y = e^u$.

Therefore $y = g(u)$ so y is a function of u which is itself a function of x.

Hence y is a function of a function of x so to find $\dfrac{dy}{dx}$ we must use the chain rule:

$$u = x^2 + x - 3 \Rightarrow \frac{du}{dx} = (2x + 1)$$

$$y = e^u \Rightarrow \frac{dy}{du} = e^u$$

Method notes

a) 5 is unaffected by the differentiation as it is a constant multiplier.

To differentiate $y = e^{3x}$ use the general result:

if $y = e^{ax}$ where a is a constant then $\dfrac{dy}{dx} = ae^{ax}$

with $a = 3$

b) to differentiate

$y = e^{2x}$ use $a = 2$ in the general formula.

The chain rule states: $\dfrac{dy}{dx} = \dfrac{dy}{du} \times \dfrac{du}{dx}$ therefore:

$$\frac{dy}{dx} = e^u \times (2x + 1) = (2x + 1)\, e^{x^2+x-3}$$

d) $y = \dfrac{e^{4x}}{x^2}$ is a quotient of two functions of x:

$$u = e^{4x} \Rightarrow \frac{du}{dx} = 4e^{4x}$$

$$v = x^2 \Rightarrow \frac{dv}{dx} = 2x$$

The quotient rule states:

$$\frac{dy}{dx} = \frac{v\dfrac{du}{dx} - u\dfrac{dv}{dx}}{v^2}$$

$$\Rightarrow \frac{dy}{dx} = \frac{x^2 \times 4e^{4x} - e^{4x} \times 2x}{x^4} = \frac{2e^{4x}(2x - 1)}{x^3}$$

Differentiation of logarithmic functions

The logarithmic function $y = \ln x$ is the inverse function of the exponential function e^x. If $y = \ln x$ then $x = e^y$.

To find $\dfrac{dy}{dx}$ when $y = \ln x$ we start with the equivalent statement $x = e^y$ as we know how to differentiate an exponential function.

If $x = e^y$ then differentiating with respect to y gives:

$$\frac{dx}{dy} = e^y.$$

We know that

$$\frac{dy}{dx} = \frac{1}{\left(\dfrac{dx}{dy}\right)} \text{ therefore } \frac{dy}{dx} = \frac{1}{e^y}$$

but $e^y = x$ therefore $\dfrac{dy}{dx} = \dfrac{1}{x}$

Hence $y = \ln x$ then $\dfrac{dy}{dx} = \dfrac{1}{x}$

This is the rule of differentiation for a natural logarithmic function.

We can now extend this further to the differentiation of composite functions involving the natural logarithm.

Example
Differentiate $y = \ln 5x^2$ with respect to x.

Answer
$y = \ln 5x^2$ is a function of a function of x.

Let $u = 5x^2$ so $\dfrac{du}{dx} = 10x$ and

let $y = \ln u$ so $\dfrac{dy}{du} = \dfrac{1}{u}$

Using the chain rule $\dfrac{dy}{dx} = \dfrac{dy}{du} \times \dfrac{du}{dx}$

$\Rightarrow \dfrac{dy}{dx} = \dfrac{1}{u} \times (10x) = \dfrac{10x}{u}$

but $u = 5x^2$ so substituting u in the equation above gives $\dfrac{dy}{dx} = \dfrac{10x}{5x^2} = \dfrac{2}{x}$

If we had used the function notation with $f(x) = 5x^2$ then $f'(x) = 10x$
We were given $y = \ln 5x^2$ therefore $y = \ln f(x)$ would be the function we had to differentiate. From our previous answer:

$\dfrac{dy}{dx} = \dfrac{10x}{5x^2}$ therefore substituting for $f(x)$ and $f'(x)$ gives $\dfrac{dy}{dx} = \dfrac{f'(x)}{f(x)}$

If $y = \ln f(x)$ then $\dfrac{dy}{dx} = \dfrac{f'(x)}{f(x)}$

This is the rule of differentiation for composite natural logarithmic functions.

Exam tips
You need to remember this formula for differentiating natural logarithm functions as it is not in the examination booklet.

Example
Differentiate the following with respect to x:

a) $y = \ln 5x$

b) $y = \ln(x^2 - 3x + 4)$

c) $y = (x^2 + 1) \ln x$

d) $y = e^2x \ln(3x)$

Answer
a) $y = \ln 5x$ is a composite function of x.

Let $f(x) = 5x \Rightarrow f'(x) = 5$

Using $\dfrac{dy}{dx} = \dfrac{f'(x)}{f(x)} \Rightarrow \dfrac{dy}{dx} = \dfrac{5}{5x} = \dfrac{1}{x}$

b) $y = \ln(x^2 - 3x + 4)$ is a composite function of x.

Let $f(x) = x^2 - 3x + 4$

$\Rightarrow f'(x) = 2x - 3$

Method notes
In (a) you could rewrite y using the rule of logs,
$y = \ln 5x$
$= \ln 5 + \ln x$, so
$\dfrac{dy}{dx} = 0 + \dfrac{1}{x} = \dfrac{1}{x}$

☞ **Continued on the next page**

Using $\dfrac{dy}{dx} = \dfrac{f'(x)}{f(x)}$

$\Rightarrow \dfrac{dy}{dx} = \dfrac{(2x - 3)}{x^2 - 3x + 4}$

c) $y = (x^2 + 1) \ln x$ is a product of two functions of x.

Let $u = (x^2 + 1) \Rightarrow \dfrac{du}{dx} = 2x$

and $v = \ln x \Rightarrow \dfrac{dv}{dx} = \dfrac{1}{x}$

Using $\dfrac{dy}{dx} = \left(\dfrac{du}{dx} \times v \right) + \left(u \times \dfrac{dv}{dx} \right)$

$\Rightarrow \dfrac{dy}{dx} = 2x \ln x + (x^2 + 1) \times \dfrac{1}{x} = 2x \ln x + \dfrac{x^2 + 1}{x}$

d) $y = e^{2x} \ln(3x)$ is a product of two functions of x.

Let $u = e^{2x} \Rightarrow \dfrac{du}{dx} = 2 e^{2x}$

and $v = \ln(3x) \Rightarrow \dfrac{dv}{dx} = \dfrac{3}{3x} = \dfrac{1}{x}$

Using $\dfrac{dy}{dx} = \left(\dfrac{du}{dx} \times v \right) + \left(u \times \dfrac{dv}{dx} \right)$

$\Rightarrow \dfrac{dy}{dx} = 2e^{2x} \ln(3x) + e^{2x} \times \dfrac{1}{x} = e^{2x}\left(2 \ln(3x) + \dfrac{1}{x} \right)$

Essential notes

If $y = e^{ax}$ where a is a constant then $\dfrac{dy}{dx} = a e^{ax}$.

Stop and think 1

Find $\dfrac{dy}{dx}$ if $y = \dfrac{e^{2x}}{\ln(3x)}$

Differentiation of trigonometric functions

To differentiate the basic trigonometric functions we start with $y = \sin x$ and then we can use the rules of differentiation to find the derivative of the other two trigonometric functions: $\cos x$ and $\tan x$.

Derivative of $\sin x$

Start with the formal definition of the gradient function as:

$f'(x)$ is the limit of $\dfrac{f(x + h) - f(x)}{h}$ as $h \to 0$

With $f(x) = \sin x$

$\dfrac{f(x + h) - f(x)}{h} = \dfrac{\sin(x + h) - \sin x}{h}$

Use the compound angle formula to expand $\sin(x + h)$:

$$\frac{\sin(x + h) - \sin x}{h} = \frac{\sin x \cos h + \cos x \sin h - \sin x}{h}$$

$$\frac{\sin(x + h) - \sin x}{h} = \sin x \left(\frac{\cos h - 1}{h}\right) + \cos x \left(\frac{\sin h}{h}\right)$$

The following table shows what happens to $\dfrac{\cos h - 1}{h}$ and $\dfrac{\sin h}{h}$ as h gets smaller using the **calculator in radians mode**.

h	0.1	0.01	0.001	0.0001
$\dfrac{\cos h - 1}{h}$	-0.049958	-0.005	-0.0005	-0.00005
$\dfrac{\sin h}{h}$	0.998334	0.999983	0.9999998	0.99999998

The table values suggest that the expression $\dfrac{\cos h - 1}{h}$ tends towards the value 0 as h tends to 0

The table also suggests that the expression $\dfrac{\sin h}{h}$ tends towards the value 1 as h tends to 0

This means that $\sin x \left(\dfrac{\cos h - 1}{h}\right) + \cos x \left(\dfrac{\sin h}{h}\right)$ tends towards $\cos x$ as $h \to 0$

If $f(x) = \sin x$ then $f'(x) = \cos x$ or

if $y = \sin x$ then $\dfrac{dy}{dx} = \cos x$

This is the rule of differentiation for the function

$y = f(x) = \sin x$.

Essential notes

The formal definition of the gradient function was covered in Core 2.

The compound angle formula for $\sin A - \sin B$ was covered in chapter 2.

Use $A = (x + h)$ and $B = x$.

Essential notes

It is essential to work in radians mode when evaluating the basic trigonometric ratios of small angles. In radians mode if h is small $\sin h \cong h$ and $\cos h \cong 1$ If you used degree mode then $\sin 1° \neq 1$

Essential notes

It is important to remember that to use this rule x **must be measured in radians.**

Example

Differentiate the following with respect to x:

a) $y = \sin 5x$

b) $y = \sin^2 x$

c) $y = x^2 \sin x$

Answer

a) $y = \sin 5x$ is a function of a function of x.

Let $u = 5x \Rightarrow \dfrac{du}{dx} = 5$ and $y = \sin u \Rightarrow \dfrac{dy}{du} = \cos u$

Using the chain rule $\dfrac{dy}{dx} = \dfrac{dy}{du} \times \dfrac{du}{dx} \Rightarrow \dfrac{dy}{dx} = 5 \cos u$

but $u = 5x \Rightarrow \dfrac{dy}{dx} = 5 \cos 5x$

Method notes

a) and b)

To differentiate a function of a function (or composite function) use the chain rule.

◄ Continued on the next page

Method notes

c) To differentiate a product of two functions of x use the product rule.

b) $y = \sin^2 x = (\sin x)^2$ is a function of a function of x.

Let $u = \sin x \Rightarrow \dfrac{du}{dx} = \cos x$ and $y = u^2 \Rightarrow \dfrac{dy}{du} = 2u$

Using the chain rule $\dfrac{dy}{dx} = \dfrac{dy}{du} \times \dfrac{du}{dx} \Rightarrow \dfrac{dy}{dx} = 2u \times \cos x$

but $u = \sin x \Rightarrow \dfrac{dy}{dx} = 2 \sin x \cos x$

c) $y = x^2 \sin x$ is a product or two functions of x.

Let $u = x^2 \Rightarrow \dfrac{du}{dx} = 2x$ and $v = \sin x \Rightarrow \dfrac{dv}{dx} = \cos x$

Using the product rule $\dfrac{dy}{dx} = \left(\dfrac{du}{dx} \times v \right) + \left(u \times \dfrac{dv}{dx} \right)$

$\Rightarrow \dfrac{dy}{dx} = (2x \sin x) + (x^2 \cos x)$

The answer to a) in the example above illustrates the general result that if

$y = \sin kx$ then $\dfrac{dy}{dx} = k \cos kx$ where k is a constant.

Derivative of cos x

If $y = \cos x$ then $y = \sin \left(\dfrac{\pi}{2} - x \right)$.

$y = \sin \left(\dfrac{\pi}{2} - x \right)$ is a function of a function (or composite function) of x.

Let $u = \left(\dfrac{\pi}{2} - x \right) \Rightarrow \dfrac{du}{dx} = -1 \left(\dfrac{\pi}{2} \text{ is a constant} \right)$

Essential notes

In any triangle ABC which is right angled at B then

$\cos C = \sin A$ and

$\angle A + \angle C = \dfrac{\pi}{2}$ (or 90°)

$\angle A = \left(\dfrac{\pi}{2} - C \right)$

$\cos C = \sin A$

$\cos C = \sin \left(\dfrac{\pi}{2} - C \right)$

In this proof let $C = x$ so:

$\cos x = \sin \left(\dfrac{\pi}{2} - x \right)$.

Or by using the compound angle formula $\sin \left(\dfrac{\pi}{2} - x \right)$

$= \sin \dfrac{\pi}{2} \cos x - \cos \dfrac{\pi}{2} \sin x$

and $y = \sin u \Rightarrow \dfrac{dy}{du} = \cos u$

Use the chain rule:

To differentiate $y = \sin \left(\dfrac{\pi}{2} - x \right)$ use the chain rule

$\dfrac{dy}{dx} = \dfrac{dy}{du} \times \dfrac{du}{dx}$

$\Rightarrow \dfrac{dy}{dx} = \cos u \times (-1) = -\cos u$

but $u = \left(\dfrac{\pi}{2} - x \right)$

$\Rightarrow \dfrac{dy}{dx} = -\cos \left(\dfrac{\pi}{2} - x \right) = -\sin x$

Therefore:

If $y = \cos x$ then $\dfrac{dy}{dx} = -\sin x$

or $f(x) = \cos x$ then $f'(x) = -\sin x$

This is the rule of differentiation for the function $y = f(x) = \cos x$. Remember the minus sign in the answer!

Example

Differentiate the following with respect to x:

a) $y = \cos 3x$ b) $y = x \cos x$

Answer

a) $y = \cos 3x$ is a composite function of x.

Let $u = 3x \Rightarrow \dfrac{du}{dx} = 3$

and $y = \cos u \Rightarrow \dfrac{dy}{du} = -\sin u$

Using the chain rule: $\dfrac{dy}{dx} = \dfrac{dy}{du} \times \dfrac{du}{dx}$

$\Rightarrow \dfrac{dy}{dx} = -\sin u \times 3 = -3\sin u$

but $u = 3x$ so $\dfrac{dy}{dx} = -3\sin 3x$

b) $y = x \cos x$ is a product of two functions of x.

Let the first function be $u = x \Rightarrow \dfrac{du}{dx} = 1$ and the second function be

$v - \cos x \Rightarrow \dfrac{dv}{dx} - -\sin x$

Using the product rule $\dfrac{dy}{dx} = \left(\dfrac{du}{dx} \times v \right) + \left(u \times \dfrac{dv}{dx} \right)$

$\Rightarrow \dfrac{dy}{dx} = (1 \times \cos x) + (1x \times (-\sin x))$

$\Rightarrow \dfrac{dy}{dx} = \cos x - x \sin x$

The answer to a) in the example above illustrates the general result if

$y = \cos kx$ then $\dfrac{dy}{dx} = -k\sin kx$ where k is a constant.

Method notes

In a) and b) remember the negative sign in the derivative of $\cos x$.

Stop and think 2

Differentiate the following with respect to x:

a) $f(x) = x^2 \cos x$

b) $f(x) = \cos^2 3x$

Derivative of $\tan x$

If $y = \tan x$ then $y = \dfrac{\sin x}{\cos x}$ which is a quotient of two functions of x.

u is the 'numerator' function: $u = \sin x \Rightarrow \dfrac{du}{dx} = \cos x$

v is the 'denominator' function: $v = \cos x \Rightarrow \dfrac{dv}{dx} = -\sin x$

Using the quotient rule $\dfrac{dy}{dx} = \dfrac{v\dfrac{du}{dx} - u\dfrac{dv}{dx}}{v^2}$

Essential notes

The basic trigonometric identities

$$\tan x = \frac{\sin x}{\cos x}$$

$$\sin^2 x + \cos^2 x \equiv 1$$

were explained in Core 2 and

$$\sec x = \frac{1}{\cos x}$$ was covered in chapter 2 of this book.

Remember that
$$\sec^2 x = (\sec x)^2$$

$$\frac{dy}{dx} = \frac{\cos x \times \cos x - \sin x \times (-\sin x)}{\cos^2 x}$$

$$= \frac{\cos^2 x + \sin^2 x}{\cos^2 x} = \frac{1}{\cos^2 x} \Rightarrow \frac{dy}{dx} = \sec^2 x$$

Therefore:

if $y = \tan x$ then $\dfrac{dy}{dx} = \sec^2 x$ or

if $f(x) = \tan x$ then $f'(x) = \sec^2 x$

This is the rule of differentiation for the function $y = f(x) = \tan x$ and more generally if $y = \tan kx$ then $\dfrac{dy}{dx} = k \sec^2 kx$ where k is a constant.

Example
Differentiate $y = x^2 \tan x$ with respect to x.

Answer
$y = x^2 \tan x$ is a product of two functions of x.

Let $u = x^2 \Rightarrow \dfrac{du}{dx} = 2x$ and

let $v = \tan x \Rightarrow \dfrac{dv}{dx} = \sec^2 x$

Using the product rule $\dfrac{dy}{dx} = \left(\dfrac{du}{dx} \times v\right) + \left(u \times \dfrac{dv}{dx}\right)$

$$\Rightarrow \qquad \frac{dy}{dx} = 2x \tan x + x^2 \sec^2 x$$

Summary
Putting the three rules for trigonometric differentiation together:

If $y = \sin kx$ then $\dfrac{dy}{dx} = k \cos kx$ where k is a constant.

If $y = \cos kx$ then $\dfrac{dy}{dx} = -k \sin kx$ where k is a constant.

If $y = \tan kx$ then $\dfrac{dy}{dx} = k \sec^2 kx$ where k is a constant.

To apply these rules for differentiating trigonometric functions you must use x measured in radians and not in degrees.

Be careful with the derivative of $\cos x$. Remember the negative sign!

Exam tips

You must learn these formulae for differentiating trigonometric functions as they are not in the examination booklet.

Example
a) Find the coordinates of the stationary points of the graph of
$y = e^{2x} \cos x$ for $0 \leq x \leq \pi$.

b) Use the second derivative method to classify the stationary points.

Answer

a) $y = e^{2x}\cos x \Rightarrow \dfrac{dy}{dx} = 2e^{2x}\cos x + e^{2x}(-\sin x)$

At stationary points $\dfrac{dy}{dx} = 0 \Rightarrow 2e^{2x}\cos x + e^{2x}(-\sin x) = 0$

$\Rightarrow e^{2x}(2\cos x - \sin x) = 0$

$\Rightarrow 2\cos x - \sin x = 0$ (e^{2x} cannot be equal to 0 as the curve $y = e^{2x}$ never crosses the x-axis)

$\Rightarrow$ if $2\cos x - \sin x = 0$ then $\dfrac{\sin x}{\cos x} = 2$

$\Rightarrow \tan x = 2$

For the range $0 \le x \le \pi$ $\tan x = 2$ for $x = 1.1071$

$x = 1.1071$ gives $y = e^{2x}\cos x = 4.094$

Therefore the coordinates of the stationary point in the range $0 \le x \le \pi$ are (1.1071, 4.094).

b) $\dfrac{dy}{dx} = 2e^{2x}\cos x + e^{2x}(-\sin x)$

$\Rightarrow \dfrac{d^2y}{dx^2} = 4e^{2x}\cos x + 2e^{2x}(-\sin x) + (-2e^{2x}\sin x - e^{2x}\cos x)$

$\Rightarrow \dfrac{d^2y}{dx^2} = e^{2x}(3\cos x - 4\sin x)$

From a) there is a stationary point when

$x = 1.1071 \Rightarrow \dfrac{d^2y}{dx^2} = -20.47$

As the second derivative is negative, the stationary point at (1.1071, 4.094) is a local maximum.

Method notes

a) This is a product of two functions of x:

$u = e^{2x} \Rightarrow \dfrac{du}{dx} = 2e^{2x}$ and

$v = \cos x \Rightarrow \dfrac{dv}{dx} = -\sin x.$

Method notes

Remember to put your calculator in radian mode.

$\tan x = 2 \Rightarrow x = \arctan 2$

Method notes

$2e^{2x}\cos x$ is a product of two functions of x:

$u = 2e^{2x}$
$\Rightarrow \dfrac{du}{dx} = 2(2e^{2x}) = 4e^{2x}$
$v = \cos x$
$\Rightarrow \dfrac{dv}{dx} = -\sin x$
$e^{2x}(-\sin x)$ is also a product of two functions of x:
$u = e^{2x} \Rightarrow \dfrac{du}{dx} = 2e^{2x}$
$v = -\sin x$
$\Rightarrow \dfrac{dv}{dx} = -\cos x$

Derivative of sec x, cosec x and cot x

Example
Differentiate the following reciprocal trigonometric functions with respect to x.

a) $y = \sec x$ b) $y = \operatorname{cosec} x$ c) $y = \cot x$

Method notes

Chain rule states:

$$\frac{dy}{dx} = \frac{dy}{du} \times \frac{du}{dx}$$

a) $\dfrac{\sin x}{\cos^2 x} = \dfrac{1}{\cos x} \times \dfrac{\sin x}{\cos x}$

$\qquad = \sec x \tan x$

b) $\dfrac{\cos x}{\sin^2 x} = \dfrac{1}{\sin x} \times \dfrac{\cos x}{\sin x}$

$\qquad = \text{cosec } x \cot x$

c) $\dfrac{\sec^2 x}{\tan^2 x} = \dfrac{1}{\cos^2 x} \times \dfrac{\cos^2 x}{\sin^2 x}$

$\qquad = \dfrac{1}{\sin^2 x} = \text{cosec}^2 x$

Answer

a) $y = \sec x$

$$\Rightarrow y = \frac{1}{\cos x} = (\cos x)^{-1}$$

which is a composite function of x.

Let $u = \cos x$ hence $\dfrac{du}{dx} = -\sin x$ and

let $y = (u)^{-1}$ hence $\dfrac{dy}{du} = -1(u)^{-2}$

Using the chain rule:

$$\frac{dy}{dx} = -1(\cos x)^{-2}(-\sin x)$$

$$= \frac{\sin x}{\cos^2 x} = \sec x \tan x$$

b) $y = \text{cosec } x$

$$\Rightarrow y = \frac{1}{\sin x} = (\sin x)^{-1} \text{ which is a composite function of } x.$$

Let $u = \sin x \Rightarrow \dfrac{du}{dx} = \cos x$ and

let $y = (u)^{-1}$ hence $\dfrac{dy}{du} = -1(u)^{-2}$

Using the chain rule:

$$\frac{dy}{dx} = -1(\sin x)^{-2}(\cos x)$$

$$= -\frac{\cos x}{\sin^2 x}$$

$$= -\text{cosec } x \cot x$$

c) $y = \cot x$

$$\Rightarrow y = \frac{1}{\tan x} = (\tan x)^{-1} \text{ which is a composite function of } x.$$

Let $u = \tan x \Rightarrow \dfrac{du}{dx} = \sec^2 x$ and

let $y = (u)^{-1}$ hence $\dfrac{dy}{du} = -1(u)^{-2}$

Using the chain rule:

$$\frac{dy}{dx} = -(\tan x)^{-2}(\sec x)^2$$

$$= -\left(\frac{\sec x}{\tan x}\right)^2$$

$$= -\text{cosec}^2 x$$

Stop and think answers

1. $y = \dfrac{e^{2x}}{\ln(3x)}$ is a quotient of two functions of x where the 'numerator function' $u = e^{2x}$ and the 'denominator' function $v = \ln(3x)$

$u = e^{2x} \Rightarrow \dfrac{du}{dx} = 2e^{2x}$ (using the chain rule)

and

$v = \ln(3x) \Rightarrow \dfrac{dv}{dx} = \left(\dfrac{1}{3x}\right) \times 3 = \dfrac{1}{x}$ (using the chain rule)

Applying the quotient rule for differentiation gives

$\dfrac{dy}{dx} = \dfrac{2e^{2x}\ln 3x - \left(e^{2x}\left(\dfrac{1}{x}\right)\right)}{(\ln 3x)^2} = \dfrac{e^{2x}\left(2\ln 3x - \dfrac{1}{x}\right)}{(\ln 3x)^2}$

2a) $f(x) = x^2\cos x$ is a product of two functions of x. To differentiate use the product rule with $u = x^2 \Rightarrow \dfrac{du}{dx} = 2x$ and

$v = \cos x \Rightarrow \dfrac{dv}{dx} = -\sin x$

Applying the product rule for differentiation gives:

$f'(x) = (\cos x)(2x) + (x^2)(-\sin x) = x(2\cos x - x\sin x)$

b) $f(x) = \cos^2(3x) = (\cos 3x)^2$ which is composite function of x so to differentiate use the chain rule.

Let $u = \cos 3x \Rightarrow \dfrac{du}{dx} = 3(-\sin 3x)$

and $f(x) = u^2 \Rightarrow \dfrac{d(f(x))}{du} = 2u$

Applying the chain rule gives:

$f'(x) = \dfrac{df(x)}{du} \times \dfrac{du}{dx} = (2u) \times 3(-\sin 3x) = -6u(\sin 3x) = -6(\cos 3x)(\sin 3x)$

When applying mathematics in problem solving we often need to solve the equation f(x) = 0 (or y = 0) as illustrated below.

Definition

Values of x which satisfy the equation f(x) = 0 (or y = 0) are called the **roots** of the equation f(x) = 0 (or y = 0).

Locating the roots of an equation

You are familiar with the process of finding the roots of a quadratic equation from Core 1.

For example, if $y = f(x) = x^2 - 5x + 2$ then you can solve $f(x) = x^2 - 5x + 2 = 0$ using "the formula for quadratics":

$$x = \frac{-b \pm \sqrt{b^2 - 4ac}}{2a}$$

In this case, $a = 1$, $b = -5$, $c = 2$

So the roots are

$$x = \frac{-(-5) \pm \sqrt{(-5)^2 - 4 \times 1 \times 2}}{2 \times 1}$$

$$= \frac{5 \pm \sqrt{17}}{2}$$

$$= 4.56 \text{ or } 0.44$$

The graph of $y = x^2 - 5x + 2$ is shown in Figure 5.1 below.

Fig. 5.1
The graph of $y = x^2 - 5x + 2$

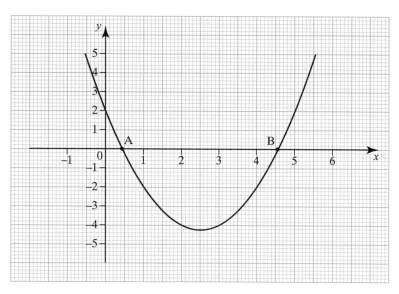

The roots of the equation are the x-coordinates of the points A and B because at these points y = 0 and from the graph of $y = x^2 - 5x + 2$ you can see that the y value changes sign as the graph crosses the x-axis at A and B.

Change of sign method

Consider a point to the left of the point A on the graph

where $x = 0.4$

$$\Rightarrow y = x^2 - 5x + 2$$
$$= (0.4)^2 - 5(0.4) + 2$$
$$= +0.16 \text{ so f}(x) \text{ (or } y) \text{ is positive to the left of the point A.}$$

Consider a point to the right of the point A on the graph

where $x = 0.5$

$$\Rightarrow y = x^2 - 5x + 2$$
$$= (0.5)^2 - 5(0.5) + 2$$
$$= -0.25 \text{ so f}(x) \text{ (or } y) \text{ is negative to the right of the point A.}$$

Since f(x) has changed sign from positive (at $x = 0.4$) to negative (at $x = 0.5$) then f(x) must $= 0$ at some point between these two x-values.

We can also see that the graph cuts the x-axis between $x = 0.4$ and $x = 0.5$ which confirms our deduction that f$(x) = 0$ between $x = 0.4$ and $x = 0.5$ hence there is a root of the equation between $x = 0.4$ and $x = 0.5$.

Taking x values to the left and right of $x = 4.56$

$x = 4.5$

$$\Rightarrow y = x^2 - 5x + 2$$
$$= (4.5)^2 - 5(4.5) + 2$$
$$= -0.25$$

$x = 4.6$

$$\Rightarrow y = x^2 - 5x + 2$$
$$= (4.6)^2 - 5(4.6) + 2$$
$$= +0.16$$

f(x) changes sign from negative to positive so f$(x) = 0$ at some point between these two x-values and as the graph cuts the x-axis between $x = 4.5$ and $x = 4.6$ we deduce that f$(x) = 0$ between $x = 4.5$ and $x = 4.6$

The graphical technique for finding where the sign of f(x) (or y) changes is useful for locating the roots of an equation which will not easily factorise.

Essential notes

'Interval' means within the range of values given for x.

Method notes

Make sure that your calculator is in radian mode since x is given in radians.

a) Substitute for $x = 2.2$ and $x = 2.3$

The change of sign implies that $f(x)$ is zero between $x = 2.2$ and $x = 2.3$

b) We can stop creating the table when there is a sign change and we have the required degree of accuracy required – in this case to 2 decimal places.

Example

a) Show that $f(x) = \sin x - \ln x$ has a root in the interval
$$2.2^c < x < 2.3^c$$

b) Find a root of the equation $\sin x - \ln x = 0$ correct to 2 decimal places.

Answer

a) $f(2.2) = \sin (2.2) - \ln (2.2) = +0.02$

$f(2.3) = \sin (2.3) - \ln (2.3) = -0.087$

The sign change in the function values confirms that there is a root in the interval $2.2 < x < 2.3$

b) From a) we know that there is a root between $x = 2.2$ and $x = 2.3$

To proceed further and obtain more accuracy for x we take values of the function between $x = 2.2$ and $x = 2.3$ and tabulate the results:

x	2.21	2.22
$f(x)$	0.0096	−0.0009

The sign change for $f(x)$ shows there is a root between $x = 2.21$ and $x = 2.22$. The $f(x)$ values show that the root is nearer 2.22 than 2.21. We deduce that correct to 2 decimal places, $x = 2.22^c$

The change of sign method relies on the continuity of the graph of $f(x)$ within the interval you are considering. Figure 5.2 below shows the graph of the function $y = f(x) = \dfrac{1}{x - 1}$

Fig. 5.2

The graph of $y = f(x) = \dfrac{1}{x - 1}$

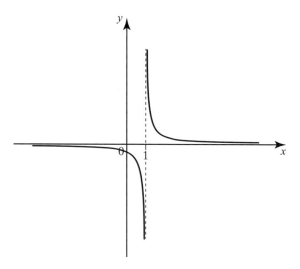

The line $x = 1$ is an asymptote to this curve because

$y \rightarrow \infty$ when $x = 1$

For a point to the left of $x = 1$ $y < 0$ and for a point on the right of $x = 1$
$y > 0$ and you can see quite clearly that there is no root of the equation

$\dfrac{1}{x - 1} = 0$ at $x = 1$

There is a discontinuity at $x = 1$

When using the method to show that there is a root between two values
of x, it important to state that **the function is continuous** between the
two values of x.

Method notes

If $x = 1$

$y = \dfrac{1}{x - 1} = \dfrac{1}{0} \rightarrow \infty$

An asymptote is a line along
which a curve approaches ∞.

Example
a) Sketch a graph of the functions $y = 2x^2 + x$ and $y = 2e^x$

b) Show that there is one solution of the equation $f(x) = 2x^2 + x - 2e^x = 0$

c) Find an interval $[a, b]$ within which the root of the equation
$2x^2 + x - 2e^x = 0$ lies where a and b are correct to one decimal place.

Essential notes

Within the interval $[a, b]$
means $a \leq x \leq b$.

Answer
a)

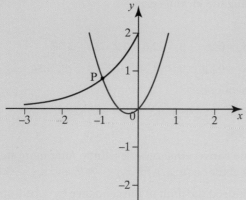

Fig. 5.3
Graphs of $y = 2x^2 + x$ (in red) and
$y = 2e^x$ (in blue).

Method notes

The graphs are sketches from
your knowledge of the graphs
of quadratic functions and
exponential functions.

$y = 2x^2 + x$ is a quadratic
function with roots at $x = 0$
and $x = -0.5$

$y = 2e^x$ is an increasing
function with y-intercept
$(0, 2)$. Exponential and
increasing functions were
discussed in Core 2.

b) The two graphs $y = 2x^2 + x$ and $y = 2e^x$ meet when the x- and y-values
are equal on both graphs. This means that

$2x^2 + x = 2e^x \Rightarrow 2x^2 + x - 2e^x = 0$

So the solution of this equation is where the two graphs meet. The
curves meet at only one point, P, so there is only one value of x that
satisfies $f(x) = 2x^2 + x - 2e^x = 0$

Continued on the next page

5 Numerical methods

c) The function f(x) is a continuous function so we can use the change of sign method to locate a root.

x	−0.5	−0.6	−0.7	−0.8	−0.9	−1
f(x)	−1.21	−0.98	−0.71	−0.42	−0.093	+0.26

Figure 5.3 also shows that the x-coordinates of the point P where the two graphs meet are between −0.5 and −1. Take other x-values up to $x = -1$ and complete a table of values until there is a sign change for f(x).

f(x) changes sign between $x = -0.9$ and $x = -1$ so there is a root in the interval [−1, −0.9].

Approximate solutions of equations

Iterative methods

An **iterative method** in mathematics is a process of continually repeating a formula to produce a sequence of values. You need to be given a starting value of x which has the symbol x_0 to begin generating the sequence with the given formula. For example, given the starting value $x_0 = 1$ and the formula $x_{n+1} = 0.2x_n - 3$ using the formula gives:

$x_1 = 0.2x_0 - 3 = 0.2 - 3 = -2.8$

If $n = 1$ the formula gives $x_2 = 0.2x_1 - 3 = -2.8 \times 0.2 - 3 = -3.56$

If $n = 2$ the formula gives $x_3 = 0.2x_2 - 3 = -3.56 \times 0.2 - 3 = -3.712$

Continuing this gives the sequence:

$x_0 = 1 \quad x_1 = -2.8 \quad x_2 = -3.56 \quad x_3 = -3.712 \quad x_4 = -3.7424$

$x_5 = -3.74848 \quad x_6 = -3.74970$

The sequence appears to be **converging** towards a limit. After more steps a calculator gives the values of the sequence for $x_{14} = -3.75$ and $x_{15} = -3.75$

You can verify that this is indeed the limit of the sequence by substituting $x_n = -3.75$ into the iterative formula $x_{n+1} = 0.2\,x_n - 3$

$x_{n+1} = 0.2 \times -3.75 - 3 = -3.75$

so as x_n and x_{n+1} both have the value −3.75 we can say that the sequence has reached a limit of −3.75

Each value of x_n is called an **iteration** and we say that the **iterative formula** $x_{n+1} = 0.2x_n - 3$ leads to a **sequence of iterations**. This sequence of iterations is

1   −2.8   −3.56   −3.712   −3.7424   −3.74848   −3.74970

which converges.

Not all iterative formulae converge. Take for example, the formula
$x_{n+1} = 2x_n - 3$ with $x_0 = 1$

The sequence of iterations in this case is

1   −1   −5   −13   −29   −61   −125   −253   −509 …

This sequence **diverges**.

Example
Find one root of the equation $x^3 - 5x + 1 = 0$ using an iterative method.

Answer
Step 1: Sketch the graph of $y = x^3 - 5x + 1$ as shown in Figure 5.4 below.

Step 2: Because the graph crosses the x-axis at three points there are roots in the intervals

[−3, −2]; [0, 1] and [2, 3].

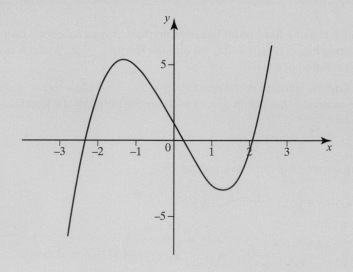

Fig. 5.4
Graph of $y = x^3 - 5x + 1$

Step 3: Rearrange the equation $x^3 - 5x + 1 = 0$ so that we have x

in terms of x^3: $x^3 + 1 = 5x \Rightarrow x = \dfrac{x^3 + 1}{5}$.

Step 4: Use the answer in step 3 to state the iterative formula

$$x_{n+1} = \frac{x_n^3 + 1}{5}$$

Essential notes

There will often be several possible rearrangements of the given equation.

Continued on the next page

Method notes

This sequence of iterations is best carried out using the 'ANS' button on your calculator.

Type in '0=' and then $(ANS^3 + 1) \div 5$

Every time you press '=' the next value of the sequence is given.

Step 5: Use the interval [0, 1] from step 2 with $x_0 = 0$ as the starting point of the formula in step 4 to generate the next term:

$$x_1 = \frac{x_0^3 + 1}{5} = \frac{1}{5} = 0.2$$

Step 6: Continue to generate the sequence using the iterative formula

$$x_2 = \frac{x_1^2 + 1}{5} = \frac{0.2^3 + 1}{5} = 0.2016$$

$x_3 = 0.20163871$ $x_4 = 0.20163965$

$x_5 = 0.20163968$ $x_6 = 0.20163968$ (to 8 decimal places).

Step 7: Substitute $x = 0.20163968$ into $f(x) = x^3 - 5x + 1$ giving

$f(0.20163968) = 0.201639683 - 5 \times 0.20163968 + 1 = -2 \times 10^{-8}$

which is close to zero. The sequence $x_0, x_1, x_2, x_3, x_4, x_5, x_6, \ldots$ converges to 0.20163968 which is therefore one root of the equation $x^3 - 5x + 1 = 0$

Essential notes

Any letter can be used to represent a function. In this case $g(x_n)$ is a function in terms of x.

This is called a **fixed point iterative method.** It uses a formula based on rearranging an equation $f(x) = 0$ into the form $x_{n+1} = g(x_n)$ which converges to a solution of $f(x) = 0$

To find the solution of the equation $x^3 - 5x + 1 = 0$ close to $x = -3$ we use the interval [−3, −2] with $x_0 = -3$ as a starting point for the iteration formula

$$x_{n+1} = \frac{x_n^3 + 1}{5}$$

The sequence of iterations is given by:

$$x_0 = -3 \qquad x_1 = \frac{-3^3 + 1}{5} = -5.2$$

$$x_2 = \frac{(-5.2)^3 + 1}{5} = -27.816 \qquad x_4 = -4304.213935 \text{ and so on.}$$

In this case the sequence of iterations **diverges**. Therefore the iterative formula $x_{n+1} = \frac{x_n^3 + 1}{5}$ cannot be used to find the solution of the equation

$x^3 - 5x + 1 = 0$ close to $x = -3$

Exam tips

This property $|g'(x_0)| < 1$ for convergence is not required for the examination.

A sequence of the form $x_{n+1} = g(x_n)$ will only converge to a root provided $|g'(x_0)| < 1$.

This illustrates that when using the fixed point iterative method the iterative formula and starting value must be chosen with care.

Example

a) Show that the equation $x^3 - 5x^2 + 7 = 0$ has a solution in the interval [1, 2].

b) Use the iteration formula $x_{n+1} = \dfrac{-7}{x_n(x_n - 5)}$ to find a solution correct to two decimal places.

Method notes

This is a cubic function which has no asymptotes so it is continuous.

Given the interval [1, 2] we use $x = 1$ and $x = 2$ to show a change of sign for $f(x)$.

Answer

a) Given $f(x) = x^3 - 5x^2 + 7$ (which is a continuous function) and substituting for x:

$x = 1 \Rightarrow f(1) = 1^3 - 5(1)^2 + 7 = +3$

$x = 2 \Rightarrow f(2) = 2^3 - 5(2)^2 + 7 = -5$

This shows that $f(x)$ changes sign between $x = 1$ and $x = 2$ so there is a root in the interval [1, 2].

b) Choose $x_0 = 1$ as the starting point for the given iteration

$$x_1 = \frac{-7}{x_0(x_0 - 5)} = \frac{-7}{1(1 - 5)} = \frac{-7}{-4} = 1.75$$

$$x_2 = \frac{-7}{x_1(x_1 - 5)} = \frac{-7}{1.75(1.75 - 5)} = 1.2308$$

similarly:

$x_3 = 1.5089$ $x_4 = 1.3288$ $x_5 = 1.4349$ $x_6 = 1.3684$

$x_7 = 1.4086$ $x_8 = 1.3837$ $x_9 = 1.3989$ $x_{10} = 1.3895$

$x_{11} = 1.3953$ $x_{12} = 1.3918$ $x_{13} = 1.3939$ $x_{14} = 1.3926\ldots$

The sequence is converging to 1.39 correct to 2 decimal places.

$x = 1.385 \Rightarrow f(1.385) = 1.385^3 - 5(1.385)^2 + 7 = +0.066$

$x = 1.395 \Rightarrow f(1.395) = 1.395^3 - 5(1.395)^2 + 7 = -0.015$

The sign change confirms the existence of the root in [1.385, 1.395].

Method notes

This sequence of iterations is best carried out using the 'ANS' button on your calculator.

Method notes

To check that the answer is correct to **2 decimal places** show that $f(x)$ changes sign between $x = 1.385$ and $x = 1.395$.

Questions

You may use a calculator.

1. The function f is defined by f: $x \mapsto \dfrac{x}{x^2 - 4} - \dfrac{1}{x + 2}$, $x > 2$

    a) Show that $f(x) = \dfrac{2}{x^2 - 4}$ (3)

    b) Find the range of f.   (2)

    c) Find the inverse function $f^{-1}(x)$, stating its domain.   (4)

    The function g is defined by

    g: $x \mapsto \ln(1 - 2x)$, $x < \dfrac{1}{2}$

    d) Find the exact value of gf(3).   (2)

2. A curve C has equation $y = x^2 e^x$.

    a) Find the coordinates of the stationary points of the curve C.   (6)

    b) Find $\dfrac{d^2 y}{dx^2}$ (2)

    c) Classify each stationary point of the curve C.   (2)

3. Differentiate with respect to x.

    a) $3 \cos^2 x + \sec 2x$ (3)

    b) $\{x + \ln(3x)\}^2$ (3)

    c) $\dfrac{4x - 1}{x^2 + 1}$ (4)

4. A function is given by $f(x) = x^3 - 2x^2 - 3$

    a) Show that the equation $f(x) = 0$ has a root in the interval $[2, 3]$.   (2)

    b) Show that $f(x) = 0$ can be rearranged as

    $$x = \sqrt{\left(2x + \dfrac{3}{x}\right)}, \; x \neq 0$$ (3)

    The root is estimated using the iterative formula

    $$x_{n+1} = \sqrt{\left(2x_n + \dfrac{3}{x_n}\right)} \text{ with } x_1 = 2$$

    c) Use this iterative formula to find the values of x_2, x_3 and x_4 correct to 3 decimal places.   (3)

    d) Prove that the root is 2.486 correct to 3 decimal places.   (2)

5. a) Using the identities for $\sin(A + B)$ and $\sin(A - B)$ prove the identity

    $$\sin(A + B) \sin(A - B) \equiv \sin^2 A - \sin^2 B$$ (3)

    b) Hence, without using a calculator, evaluate $\sin 75° \sin 15°$.   (2)

    c) Solve the equation

    $$\sin\left(x + \dfrac{\pi}{3}\right) \sin\left(x - \dfrac{\pi}{3}\right) = \sin x, \text{ for } 0 \leq x \leq 2\pi.$$ (5)

6. A can of coke, at a temperature $T°C$, warms according to the formula
   $T = 20 - 16e^{-kt}$ where t is the time in seconds after the can is taken out
   of the fridge.

   a) What is the initial temperature of the coke? (1)

   b) The coke warms by 5°C during the first minute of being out
      of the fridge.

      Find the value of k correct to 3 significant figures. (3)

   c) How long will it take for the coke to reach a temperature of 15°C? (3)

7. a) Sketch on the same diagram graphs of the two functions $y = x^2 - 4a^2$

      and $y = |x - 2a|$ where a is a positive constant.

      Show, in terms of a, where each graph meets the coordinate axes. (4)

   b) Find, in terms of a, the coordinates of the points of intersection
      of the two graphs. (4)

8. a) Write $\sin x + \sqrt{3} \cos x$ in the form $R\cos(x - \alpha)$ where $R < 0$

      and $0 < \alpha < \dfrac{\pi}{2}$. (4)

   b) Solve the equation $\sin x + \sqrt{3} \cos x = \sqrt{2}$ for x in the interval
      $-\pi \le x \le \pi$, giving your answers in terms of π. (5)

Answers

1. a) Given $f(x) = \dfrac{x}{x^2 - 4} - \dfrac{1}{x + 2}$ to rewrite as asked, factorise the

   denominator $(x^2 - 4)$ by using the difference of two squares:

   $$\dfrac{x}{x^2 - 4} - \dfrac{1}{x + 2} = \dfrac{x}{(x - 2)(x + 2)} - \dfrac{1}{(x + 2)}$$ (1)

   Find the common denominator for the two fractions which is
   $(x - 2)(x + 2)$ so

   $$\dfrac{1}{(x + 2)} = \dfrac{(x - 2)}{(x - 2)(x + 2)}$$

   therefore $\dfrac{x}{(x - 2)(x + 2)} - \dfrac{1}{(x + 2)}$

   $$= \dfrac{x}{(x - 2)(x + 2)} - \dfrac{(x - 2)}{(x - 2)(x + 2)}$$

   $$= \dfrac{x - (x - 2)}{(x - 2)(x + 2)}$$

   $$= \dfrac{2}{(x - 2)(x + 2)}$$

   $$= \dfrac{2}{(x^2 - 4)}$$ (2)

b) Given that $y = f(x) = \dfrac{x}{(x^2 - 4)}$ as the domain is $x > 2$ then $(x^2 - 4) > 0$ so

$$y = \dfrac{x}{(x^2 - 4)} > 0 \qquad (1)$$

so the range of f is $\{y: y > 0\}$. $\qquad (1)$

c) Given $y = \dfrac{2}{(x^2 - 4)}$ to find the inverse rewrite this equation to

make x the subject:

$$x^2 - 4 = \dfrac{2}{y} \Rightarrow x^2 = 4 + \dfrac{2}{y} \Rightarrow x = \sqrt{4 + \dfrac{2}{y}}$$

and we are given that $x > 0$ so we take the positive square root.

Interchanging x and y in $x = \sqrt{4 + \dfrac{2}{y}}$

$$\Rightarrow y = \sqrt{4 + \dfrac{2}{x}}$$

We now have the relationship which takes y back to x so this is the

inverse function: $f^{-1}: x \mapsto \sqrt{4 + \dfrac{2}{x}}$ $\qquad (3)$

For this to be defined we must take $x \neq 0$ otherwise $\dfrac{2}{x}$ is infinite so

the domain of f^{-1} is $\{x: x > 0\}$ $\qquad (1)$

d) Given $f(x) = y = \dfrac{2}{(x^2 - 4)}$ then $f(3) = \dfrac{2}{(3^2 - 4)} = \dfrac{2}{5}$ $\qquad (1)$

Given $g(x) = \ln(1 - 2x)$ then:

$$g\left(\dfrac{2}{5}\right) = \ln\left(1 - 2 \times \dfrac{2}{5}\right)$$

$$= \ln \dfrac{1}{5}$$

$$= \ln 5^{-1} = -\ln 5 \text{ using the rule of indices:}$$

$$gf(3) = -\ln 5 \qquad (1)$$

2. a) $y = x^2 e^x$

This is a product of two functions of x:

take $u = x^2$ which gives $\dfrac{du}{dx} = 2x$

and $v = e^x$ which gives $\dfrac{dv}{dx} = e^x$

then use the product rule:

$$\dfrac{dy}{dx} = 2xe^x + x^2e^x \qquad (2)$$

At stationary points, $\dfrac{dy}{dx} = 0$ so

$$\dfrac{dy}{dx} = 2xe^x + x^2e^x = 0$$

$$\Rightarrow (2x + x^2)e^x = 0$$

$$\Rightarrow (2x + x^2) = 0 \text{ as } e^x \text{ cannot} = 0$$

$$\Rightarrow x(2 + x) = 0$$

$$\Rightarrow x = 0 \text{ or } x = -2 \qquad (2)$$

The question asks for the coordinates of the stationary points so we must work out the y-value for each of these x-values:

if $x = 0$ in $y = x^2 e^x \Rightarrow y = 0$

if $x = -2$ in $y = x^2 e^x \Rightarrow y = 4e^{-2}$

The coordinates of the stationary points are $(0, 0)$
and $(-2, 4e^{-2})$ (2)

b) $\dfrac{dy}{dx} = 2xe^x + x^2e^x$

To find $\dfrac{d^2y}{dx^2}$ we must differentiate $\dfrac{dy}{dx}$ with respect to x. Each part of $\dfrac{dy}{dx}$ is a product of two functions of x:

for $2xe^x$, $u = 2x \Rightarrow \dfrac{du}{dx} = 2$, and $v = e^x \Rightarrow \dfrac{dv}{dx} = e^x$

for $x^2 e^x$, $u = x^2 \Rightarrow \dfrac{du}{dx} = 2x$ and $v = e^x \Rightarrow \dfrac{dv}{dx} = e^x$

Therefore $\dfrac{d^2y}{dx^2} = (2 + 2x)e^x + (2x + x^2)e^x$ (2)

c) $x = 0 \Rightarrow \dfrac{d^2y}{dx^2} = 2 > 0 \Rightarrow (0, 0)$ is a local minimum (1)

$x = -2 \Rightarrow \dfrac{d^2y}{dx^2} = -2e^{-2} < 0$

$\Rightarrow (-2, 4e^{-2})$ is a local maximum (1)

3. a) $y = 3\cos^2 x + \sec 2x$

Each of the two functions $3\cos^2 x$ and $\sec 2x$ is a composite function of x so we use the chain rule to differentiate. (1)

Rewrite $3\cos^2 x$ as $3(\cos x)^2$ and apply the chain rule to differentiate:
$\Rightarrow 3(2)(\cos x)^1 (-\sin x)$

Rewrite $\sec 2x$ as $\sec u$ where $u = 2x \Rightarrow \dfrac{du}{dx} = 2$ and apply the chain rule to differentiate:

$\Rightarrow (\sec 2x \tan 2x) \times 2$

Therefore combining the results:

$\dfrac{dy}{dx} = 3(2\cos x)(-\sin x) + 2\sec 2x \tan 2x$ (2)

$\qquad = -3\sin 2x + 2\sec 2x \tan 2x$

b) $y = [x + \ln(3x)]^2$

This is a composite function of x so we use the chain rule to differentiate. (1)

Let $u = (x + \ln(3x))$:

$\Rightarrow \dfrac{du}{dx} = 1 + \left(\dfrac{1}{3x}\right) \times 3 = 1 + \dfrac{1}{x}$

and $y = u^2 \Rightarrow \dfrac{du}{dx} = 2u$

Therefore:

$\dfrac{dy}{dx} = 2[x + \ln(3x)] \times \left(1 + \dfrac{1}{x}\right)$

$= \dfrac{2(1 + x)[x + \ln(3x)]}{x}$ (2)

c) $y = \dfrac{4x - 1}{x^2 + 1}$

This is a quotient of two functions of x so we use the quotient rule to differentiate. (1)

Let $u = (4x - 1) \Rightarrow \dfrac{du}{dx} = 4$

Let $v = (x^2 + 1) \Rightarrow \dfrac{dv}{dx} = 2x$

Therefore using the quotient rule:

$\dfrac{dy}{dx} = \dfrac{4(x^2 + 1) - (4x - 1)(2x)}{(x^2 + 1)^2}$

$= \dfrac{-4x^2 + 2x + 4}{(x^2 + 1)^2}$ (3)

4. a) Given $f(x) = x^3 - 2x^2 - 3$

$x = 2 \Rightarrow y = 2^3 - 2 \times 2^2 - 3 = -3$

$x = 3 \Rightarrow y = 3^3 - 2 \times 3^2 - 3 = +6$

$f(x)$ is a continuous function and changes sign between $x = 2$ and $x = 3$ therefore the equation $f(x) = 0$ has a root in the interval $[2, 3]$. (2)

b) Given $\qquad f(x) = 0$

$\Rightarrow \quad x^3 - 2x^2 - 3 = 0$

$\Rightarrow \qquad\qquad x^3 = 2x^2 + 3$

$\Rightarrow \qquad\qquad x^2 = \dfrac{2x^2 + 3}{x} = \dfrac{2x^2}{x} + \dfrac{3}{x}$ with $x \neq 0$

$\Rightarrow \qquad\qquad x^2 = 2x + \dfrac{3}{x}$

Taking the square root of both sides of this equation

$\Rightarrow \qquad\qquad x = \sqrt{2x + \dfrac{3}{x}}$ (3)

c) Given $x_{n+1} = \sqrt{2x_n + \dfrac{3}{x_n}}$ and $x_1 = 2$

$\Rightarrow \qquad x_2 = \sqrt{2x_1 + \dfrac{3}{x_1}}$

$\Rightarrow \qquad x_2 = 2.345$ (1)

Apply the formula again with $x_2 = 2.345$

$\Rightarrow \qquad x_3 = 2.443$ (1)

Apply the formula again with $x_3 = 2.443$

$\Rightarrow x_4 = 2.473$ (1)

d) To prove that the root is 2.486 to 3 decimal places we need to show a change of sign for function values on either side of $x = 2.486$

Start with $x = 2.4855$ (take the x-value to four decimal places to ensure accuracy for the final answer of x to three decimal places):

$f(2.4855) = -0.00072$

$f(2.4865) = +0.0079$

$f(x)$ is a continuous function and changes sign between $x = 2.4855$ and $x = 2.4865$ therefore the equation $f(x) = 0$ has a root in the interval $[2.4855, 2.4865]$. The root is 2.486 correct to 3 decimal places.   (2)

5. a) $\sin(A + B) \equiv \sin A \cos B + \cos A \sin B$ (1)

$\sin(A - B) \equiv \sin A \cos B - \cos A \sin B$ (1)

So $\sin(A+B) \sin(A-B) = (\sin A \cos B + \cos A \sin B)(\sin A \cos B - \cos A \sin B)$

$$= \sin^2 A \cos^2 B - \cos^2 A \sin^2 B$$

$$= \sin^2 A (1 - \sin^2 B) - (1 - \sin^2 A) \sin^2 B$$

$$\equiv \sin^2 A - \sin^2 B \text{ so the identity is proved.} \quad (1)$$

b) 'Hence' implies that you should use the result from (a) to answer (b)

'Without using a calculator' implies that the angles involved in answering the question will be common angles for which you know the sine and cosine.

From (a) $\sin(A + B) \sin(A - B) \equiv \sin^2 A - \sin^2 B$

Therefore comparing $\sin 75° \sin 15°$ with $\sin(A + B) \sin(A - B)$:

     let $A + B = 75°$ (1)

and let $A - B = 15°$ (2)

Adding (1) and (2)

$\Rightarrow 2A = 90°$

$\Rightarrow A = 45°$

Subtracting (1) and (2)

$\Rightarrow 2B = 60°$

$\Rightarrow B = 30°$ (1)

Use these results in (b)

$$\Rightarrow \sin 75° \sin 15° = \sin^2 45° - \sin^2 30° = \left(\frac{1}{\sqrt{2}}\right)^2 - \left(\frac{1}{2}\right)^2$$

$$= \frac{1}{2} - \frac{1}{4} = \frac{1}{4} \tag{1}$$

c) From (a) $\sin\left(x + \dfrac{\pi}{3}\right) \sin\left(x - \dfrac{\pi}{3}\right)$

$$= \sin^2 x - \sin^2 \frac{\pi}{3}$$

$$= \sin^2 x - \frac{3}{4} \tag{1}$$

Given $\sin\left(x + \dfrac{\pi}{3}\right)\sin\left(x - \dfrac{\pi}{3}\right) = \sin x$ then $\sin^2 x - \dfrac{3}{4} = \sin x$.

Multiplying throughout by 4 gives:

$4\sin^2 x - 4\sin x - 3 = 0$

$\Rightarrow (2\sin x - 3)(2\sin x + 1) = 0$

$\Rightarrow \sin x = \dfrac{3}{2}$ (reject as $\sin x < 1$) or $\sin x = -\dfrac{1}{2}$ $\qquad$ (2)

From the graph: $\sin x = -\dfrac{1}{2} \Rightarrow x = \dfrac{7\pi}{6}$ or $x = \dfrac{11\pi}{6}$ $\qquad$ (2)

Fig. 1
Graphs of $y = \sin x$ (red) and
$y = -\dfrac{1}{2}x$ (blue) showing the solutions

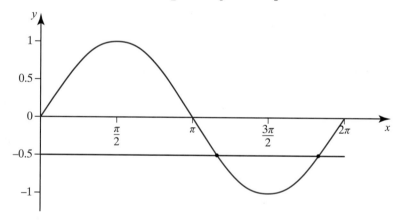

6. a) The initial (starting) temperature is when the coke is taken out of the fridge. In the formula $T = 20 - 16e^{-kt}$ t is the time in seconds after the coke is taken out of the fridge, and at this point $t = 0$

If $t = 0$, $T = 20 - 16e^0$ and as $e^0 = 1$, $T = 20 - 16 = 4$

The initial temperature is 4°C $\qquad$ (1)

b) The coke warms by 5° in the first minute so the temperature at that point is

$T = 4° + 5° = 9°$. One minute is 60 seconds (t must be in seconds), so using

$t = 60$ and $T = 9$ in $T = 20 - 16e^{-kt}$ $\qquad$ (1)

$$9 = 20 - 16e^{-60k}$$

$$\Rightarrow \quad 16e^{-60k} = 20 - 9$$

$$\Rightarrow \quad e^{-60k} = \frac{11}{16}$$

Taking ln of both sides gives

$$\ln e^{-60k} = \ln \frac{11}{16}$$

$$\Rightarrow -60k\ln e = \ln \frac{11}{16} \text{ and } \ln e = 1 \text{ (rules of logs)}$$

$$\Rightarrow \qquad k = -\frac{1}{60} \ln \frac{11}{16}$$

therefore $k = 0.00625$ correct to 3 s.f. (2)

c) To find how long it takes for the coke to reach a temperature of 15°C means we have to find t when $T = 15$ (1)

Using $T = 20 - 16e^{-kt}$

$15 = 20 - 16e^{-0.00625t}$ (using k to 3 s.f.)

$16e^{-0.00625t} = 20 - 15$

$16e^{-0.00625t} = 5$

$e^{-0.00625t} = \frac{5}{16}$

$-0.00625t\ln e = \ln \frac{5}{16}$ (rules of logs)

therefore $t = 186.104$ (seconds)

This means that it takes 3 minutes 7 seconds for the coke to reach 15°C. (2)

7. a) The graph of $y = x^2 - 4a^2$ where a is a constant is a parabola as shown in red in the diagram below. It will cross the x-axis where $y = 0$ and $x = \pm 2a$ because $x^2 = 4a^2$. (4)

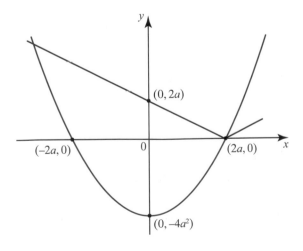

Fig. 2
Graphs of $y = x^2 - 4a^2$ (red) and $y = |(x - 2a)|$ (blue).

To sketch the graph $y = |(x - 2a)|$ draw the line $y = x - 2a$ and once the line reaches the x-axis draw its reflection above the x-axis as the modulus sign means that y can never be negative. This is shown in blue on the diagram above.

b) From the graphs $(2a, 0)$ is a point of intersection. (1)

The second point of intersection for $x < 0$ will be where the y-value on the line and the value on the curve are equal:

as $x < 0$ and $y = |(x - 2a)|$ then y would be negative so $y = -(x - 2a)$ therefore

$$x^2 - 4a^2 = y = -(x - 2a)$$

so $x^2 - 4a^2 = 2a - x$

and $x^2 + x - (4a^2 + 2a) = 0$

Factorising:

$$(x - 2a)(x + (2a + 1)) = 0$$

$\Rightarrow x = -(2a + 1)$ (reject $x = 2a$ as $x < 0$)

Substituting in $y = |(x - 2a)|$ means y is the numerical value of $x - 2a$

$\Rightarrow y = 2a - (-(2a + 1)) = 4a + 1$ (3)

The coordinates of the points of intersection of the two graphs are $(2a, 0)$ and $(-(2a + 1), (4a + 1))$

8. a) $R \cos(x - \alpha) = R (\cos x \cos \alpha + \sin x \sin \alpha)$

Let $\sqrt{3} \cos x + \sin x = R \cos x \cos \alpha + R \sin x \sin \alpha$ as given in the question with

$R > 0$ and $0 < \alpha < \dfrac{\pi}{2}$

Comparing coefficients of $\cos x$ and $\sin x$:

$R \cos \alpha = \sqrt{3}$ (1)

$R \sin \alpha = 1$ (2)

Squaring and adding (1) and (2):

$$R^2 \cos^2 \alpha + R \sin^2 \alpha = R^2 (\cos^2\alpha + \sin^2\alpha)$$

$\Rightarrow \qquad R^2 (1) = (\sqrt{3})^2 + 1^2 = 4$

$\Rightarrow \qquad R = 2$ reject $R = -2$ as $R > 0$ (2)

Dividing (2) by (1):

$$\frac{\sin \alpha}{\cos \alpha} = \tan \alpha = \frac{1}{\sqrt{3}}$$

$\Rightarrow \alpha = \dfrac{\pi}{6}$ as $0 < \alpha < \dfrac{\pi}{2}$ (2)

Therefore $\sqrt{3} \cos x + \sin x = 2 \cos\left(x - \dfrac{\pi}{6}\right)$

b) To solve $\sqrt{3}\cos x + \sin x = \sqrt{2}$, use the result from (a)

$$\Rightarrow 2\cos\left(x - \frac{\pi}{6}\right) = \sqrt{2} \qquad (1)$$

$$\Rightarrow \cos\left(x - \frac{\pi}{6}\right) = \frac{\sqrt{2}}{2} = \frac{1}{\sqrt{2}} \text{ and from common angles we know that}$$

$$\cos\frac{\pi}{4} = \frac{1}{\sqrt{2}} \qquad (2)$$

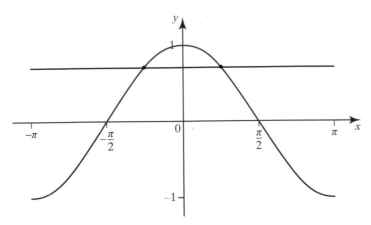

Fig. 3
Graphs of $y = \cos x$ (red) and

$y = \dfrac{\sqrt{2}}{2}$ (blue).

From the graph:

$$x - \frac{\pi}{6} = -\frac{\pi}{4} \Rightarrow x = -\frac{\pi}{12} \qquad (1)$$

$$\text{or } x - \frac{\pi}{6} = \frac{\pi}{4} \Rightarrow x = \frac{5\pi}{12} \qquad (1)$$

These are the solutions of $\sqrt{3}\cos x + \sin x = \sqrt{2}$

Set notation

$\in$	is an element of
$\notin$	is not an element of
$\{x_1, x_2, \ldots\}$	the set with elements $x_1, x_2, \ldots$
$\{x : \ldots\}$	the set of all x such that $\ldots$
$n(A)$	the number of elements in set A
$\varnothing$	the empty set
ε	the universal set
A'	the complement of the set A
$\mathbb{N}$	the set of natural numbers, $\{1, 2, 3, \ldots\}$
$\mathbb{Z}$	the set of integers, $\{0, \pm 1, \pm 2, \pm 3, \ldots\}$
$\mathbb{Z}^+$	the set of positive integers, $\{1, 2, 3, \ldots\}$
$\mathbb{Z}_n$	the set of integers modulo n, $\{0, 1, 2, \ldots, n - 1\}$
$\mathbb{Q}$	the set of rational numbers, $\left\{\dfrac{p}{q} : p \in \mathbb{Z}, q \in \mathbb{Z}^+\right\}$
$\mathbb{Q}^+$	the set of positive rational numbers, $\{x \in \mathbb{Q} : x > 0\}$
$\mathbb{Q}_0^+$	the set of positive rational numbers and zero, $\{x \in \mathbb{Q} : x \geq 0\}$
$\mathbb{R}$	the set of real numbers
$\mathbb{R}^+$	the set of positive real numbers $\{x \in \mathbb{R} : x > 0\}$
$\mathbb{R}_0^+$	the set of positive real numbers and zero, $\{x \in \mathbb{R} : x \geq 0\}$
$\mathbb{C}$	the set of complex numbers
(x, y)	the ordered pair x, y
$A \times B$	the cartesian product of sets A and B, ie $A \times B = \{(a, b) : a \in A, b \in B\}$
$\subseteq$	is a subset of
$\subset$	is a proper subset of
$\cup$	union
$\cap$	intersection
$[a, b]$	the closed interval, $\{x \in \mathbb{R} : a \leq x \leq b\}$
$[a, b), [a, b[$	the interval $\{x \in \mathbb{R} : a \leq x < b\}$
$(a, b], \,]a, b]$	the interval $\{x \in \mathbb{R} : a < x \leq b\}$
$(a, b), \,]a, b[$	the open interval $\{x \in \mathbb{R} : a < x < b\}$
$y \, R \, x$	y is related to x by the relation R
$y \sim x$	y is equivalent to x, in the context of some equivalence relation

Miscellaneous symbols

$=$	is equal to
$\neq$	is not equal to
$\equiv$	is identical to or is congruent to
$\approx$	is approximately equal to
$\cong$	is isomorphic to
$\propto$	is proportional to
$<$	is less than
$\leq$	is less than or equal to, is not greater than
$>$	is greater than
$\geq$	is greater than or equal to, is not less than
∞	infinity
$p \wedge q$	p and q
$p \vee q$	p or q (or both)
$\sim p$	not p
$p \Rightarrow q$	p implies q (if p then q)
$p \Leftarrow q$	p is implied by q (if q then p)
$p \Leftrightarrow q$	p implies and is implied by q (p is equivalent to q)
$\exists$	there exists
$\forall$	for all

Operations

$a + b$	a plus b
$a - b$	a minus b
$a \times b, ab, a.b$	a multiplied by b
$a \div b, \dfrac{a}{b}, a/b$	a divided by b
$\displaystyle\sum_{i=1}^{n} a_i$	$a_1 + a_2 + \ldots + a_n$
$\displaystyle\prod_{i=1}^{n} a_i$	$a_1 \times a_2 \times \ldots \times a_n$
$\sqrt{a}$	the positive square root of a
$|a|$	the modulus of a
$n!$	n factorial
$\dbinom{n}{r}$	the binomial coefficient $\dfrac{n!}{r!(n-r)!}$ for $n \in \mathbb{Z}^{+}$
	$\dfrac{n(n-1)\ldots(n-r+1)}{r!}$ for $n \in \mathbb{Q}$

Functions

$f(x)$	the value of the function f at x
$f : A \rightarrow B$	f is a function under which each element of set A has an image in set B
$f : x \rightarrow y$	the function f maps the element x to the element y
f^{-1}	the inverse function of the function f
$g \circ f, gf$	the composite function of f and g which is defined by $(g \circ f)(x)$ or $gf(x) = g(f(x))$
$\lim_{x \to a} f(x)$	the limit of f(x) as x tends to a
$\dfrac{dy}{dx}$	the derivative of y with respect to x
$\dfrac{d^n y}{dx^n}$	the nth derivative of y with respect to x
$f'(x), f''(x), \ldots, f(n)(x)$	the first, second, $\ldots$, nth derivatives of f(x) with respect to x
$\int y \, dx$	the indefinite integral of y with respect to x
$\int_a^b y \, dx$	the definite integral of y with respect to x between the limits $x = a$ and $x = b$

Exponential and logarithmic functions

e	base of natural logarithms
e^x, exp x	exponential function of x
$\log_a x$	logarithm to the base a of x
lin x, $\log_e x$	natural logarithm of x
lg x, $\log_{10} x$	logarithm of x to base 10

Circular and hyperbolic functions

sin, cos, tan, cosec, sec, cot	the circular functions
arcsin, arccos, arctan, arccosec, arcsec, arccot	the inverse circular functions

Vectors

$\mathbf{a}$	the vector $\mathbf{a}$
$\overrightarrow{AB}$	the vector represented in magnitude and direction by the directed line segment AB
$\hat{\mathbf{a}}$	a unit vector in the direction of $\mathbf{a}$
$\mathbf{i}, \mathbf{j}, \mathbf{k}$	unit vectors in the directions of the cartesian coordinate axes
$|\mathbf{a}|, a$	the magnitude of $\mathbf{a}$
$|\overrightarrow{AB}|, AB$	the magnitude of $\overrightarrow{AB}$
$\mathbf{a} \cdot \mathbf{b}$	the scalar product of $\mathbf{a}$ and $\mathbf{b}$.

Formulae you need to remember

Trigonometry

$$\cos^2 A + \sin^2 A \equiv 1$$

$$\sec^2 A \equiv 1 + \tan^2 A$$

$$\operatorname{cosec}^2 A \equiv 1 + \cot^2 A$$

$$\sin 2A \equiv 2 \sin A \cos A$$

$$\cos 2A \equiv \cos^2 A - \sin^2 A$$

$$\tan 2A \equiv \frac{2 \tan A}{1 - \tan^2 A}$$

Differentiation

function	derivative
$\sin kx$	$k \cos kx$
$\cos kx$	$-k \sin kx$
e^{kx}	ke^{kx}
$\ln x$	$\dfrac{1}{x}$
$f(x) + g(x)$	$f'(x) + g'(x)$
$f(x)\,g(x)$	$f'(x)\,g'(x) + f(x)\,g'(x)$
$f(g(x))$	$f'(g(x))\,g'(x)$

Formulae given in the formulae booklet

Mensuration

Surface area of sphere $= 4\pi r^2$

Area of curved surface of cone $= \pi r \times$ slant height

Arithmetic series

$$u_n = a + (n - 1)d$$

$$S_n = \frac{1}{2}n(a + l) = \frac{1}{2}n[2a + (n - 1)d]$$

Cosine rule

$$a^2 = b^2 + c^2 - 2bc \cos A$$

Binomial series

$$(a + b)^n = a^n + \binom{n}{1} a^{n-1}b + \binom{n}{2} a^{n-2}b^2 + \ldots + \binom{n}{r} a^{n-r}b^r + \ldots + b^n \quad (n \in \mathbb{N})$$

$$\text{where } \binom{n}{r} = {}^nC_r = \frac{n!}{r!(n - r)!}$$

$$(1 + x)^n = 1 + nx + \frac{n(n-1)}{1 \times 2}x^2 + \ldots + \frac{n(n-1)\ldots(n-r+1)}{1 \times 2 \times \ldots \times r}x^r + \ldots \; (|x| < 1, n \in \mathbb{R})$$

Logarithms and exponentials

$$\log_a x = \frac{\log_b x}{\log_b a}$$

Geometric Series

$$u_n = ar^{n-1}$$

$$S_n = \frac{a(1 - r^n)}{1 - r}$$

$$S_\infty = \frac{a}{1 - r} \text{ for } |r| < 1$$

Numerical integration

The trapezium rule: $\displaystyle\int_a^b y \, dx \approx \tfrac{1}{2} h\{(y_0 + y_n) + 2(y_1 + y_2 + \ldots + y_{n-1})\}$, where $h = \dfrac{b - a}{n}$

Logarithms and exponentials

$$e^{x \ln a} = a^x$$

Trigonometric identities

$$\sin(A \pm B) = \sin A \cos B \pm \cos A \sin B$$

$$\cos(A \pm B) = \cos A \cos B \mp \sin A \sin B$$

$$\tan(A \pm B) = \frac{\tan A \pm \tan B}{1 \mp \tan A \tan B} \qquad (A \pm B \neq (k + \tfrac{1}{2})\pi)$$

$$\sin A + \sin B = 2 \sin \frac{A + B}{2} \cos \frac{A - B}{2}$$

$$\sin A - \sin B = 2 \cos \frac{A + B}{2} \sin \frac{A - B}{2}$$

$$\cos A + \cos B = 2 \cos \frac{A + B}{2} \cos \frac{A - B}{2}$$

$$\cos A - \cos B = -2 \sin \frac{A + B}{2} \sin \frac{A - B}{2}$$

Differentiation

$f(x)$	$f'(x)$
$\tan kx$	$k \sec^2 kx$
$\sec x$	$\sec x \tan x$
$\cot x$	$-\operatorname{cosec}^2 x$
$\operatorname{cosec} x$	$-\operatorname{cosec} x \cot x$
$\dfrac{f(x)}{g(x)}$	$\dfrac{f'(x)\,g(x) - f(x)g'(x)}{(g(x))^2}$

Index